THE BRIDGE
TO CONSCIOUSNESS

The Bridge to Consciousness

I'm writing the bridge between science and our old and new beliefs.

David J Conway

Copyright © 2015 by David J Conway.

Library of Congress Control Number:		2015913433
ISBN:	Hardcover	978-1-5035-9562-0
	Softcover	978-1-5035-9561-3
	eBook	978-1-5035-9560-6

To order additional copies of this book, contact:
Xlibris
1-888-795-4274
www.Xlibris.com
Orders@Xlibris.com
722741

CONTENTS

My friend Taco drew this picture in my notebook and we never discussed its meaning as a picture sketches a thousand words. Within the same week a young girl called Eveline came into my life, she is a student and we chatted and I was then inspired to write this project. Eveline is my first student, she wanted to learn and she seems to be very capable of understanding, so our friendship is a trade off as she turned me into a teacher.

Introduction

This book is not only going to challenge your mind and your beliefs but it's going to challenge your very own soul! I invite you on a Journey, of not only self-discovery but also mastery. The more you know about yourself – the less there is to know full stop!

I dedicate this book to a few Legends that each had an impact on my life but sadly have passed on. Legends such as Aubrey, Adam, Mrs Brady, Moorsey, Miss Slattery, Willy, Derek, Finner, David, Billy, Dermo, 4D just to name a few. And the biggest inspiration I got was from a Lady called Florence (my Grandmother) who passed on in 2002, and a man who's name is Vincent (my Granduncle) who passed on before I was even born. Thank you, great people who lived great lives and are happy now. I salute you!

Life's a Game – you have two choices in this Life.
1) To be unconscious and to be played.
2) To be awakened and to become the Player that your very own soul not only yearns for but has a Higher Power given right to do so!

I invite you on a Journey – the choice is yours – to be a Player or to be played – this is my Gift to You!

This book is a Bridge that consists of 12 steps to walk over that Bridge and each Step consists of 12 stages. This whole book consists of text messages from myself David the Teacher to my Student Eveline, copied to notes and emailed to a Publisher. I believe that, if I can give someone back to them-selves, on the deepest level, that is my greatest gift!

There are three books that together make this a Trilogy! This is part two which is the bridge. At the start of this bridge, Step one, is book one, that consists of a subject called the Science of Numbers and in order for me to do this very personal reading for you, you need to send me your

1) Name

2) Date of Birth

3) Your email address

To my email address *djsgame@gmail.com* or better still, find me on FaceBook; The Bridge to Consciousness – book. This book the Bridge, part two of the Trilogy (honestly) is the best!

Part three is a book I wrote ten years ago and if it interests you as to how I found the answers in this book, then the book 'Life is a Game' by DJ Con, is a must read.

Step 1

The Science Of Numbers

"Evolution is the law of life. Numbers is the law of the universe. Unity is the law of God." Pythagoras.

A very brief introduction to the Science of Numbers.

For example, I was born on 19/6/1980. When that's added up as primary numbers 1+9+6+1+9+8+0. It equals 34.

Always break it down to a primary number as only 0-9 exist apart from master numbers 11, 22 and 33. 3+4=7. I'm a 7!

Inside that 7 – there are peak influences. For example behind that 7, I was a 7 again for 29 years of my life. Why 29? Well 7 (life path number) minus 36 (four cycles of 9) equal 29. At the age of 29, I reached a peak of so many years on an inward mystical journey of understanding the rules, behind the scenes and influences of life. After that I got a house, had a child, set up a company and later crashed and burned, went sleep walking for a year. I had anxiety attacks and couldn't breathe, (anxiety is cause by mind in the future, depression is the mind in the past and the answer is the Now). Because this trilogy was in the back of my mind and I felt I was not the person to do it but then I looked back at numbers and as numbers changed, my energy changed. I'm slowly becoming the person to finish this trilogy once and for all. As been a 7 and on an emotional roller coaster for so long, once that stopped, I didn't even know who I was anymore. But over time I moved from an inward 7 to a 1. I'm to be on a 1 from ages 30-39, a 1 means self-expression. This Bridge is I, self expressing myself and I am beside myself here reliving the 29 years of the 7 influence through previous experience and at the same time, I'm on the 1 and self expressing myself.

There is no number that is good and another that is bad, it's all good, bad and indifferent, it is everything and so you are too. It just is, what it is. Numbers show your deep, real self, to yourself.

In this book nothing is left out, a conversation took place and I can't hide any of that, you may read and think I'm just too honest but in my journey I have lost so many friends from states of mind they fell into and couldn't get out of and it always drove me to find the answers and now I have. I'm too late for a lot of my friends but I dedicate this knowledge to them, on their behalf and I'm just sad that this information was not around before I lost them.

"Hi Eveline, we had an interesting chat the other night and I started something with the Science of Numbers and told you that I'd send you this information, as I like to finish what I started.

So here goes. (I sent Eveline information from the Science of Numbers, her Life Path and Peak numbers and the Day influence and the 9year cycles). If you get through all that, I don't usually ever tell anyone that amount of information in one go but I think you can handle it. Okay, once through all that, you're about 60% down the rabbit hole with the science of Numbers and that's only 10% of my life's experiences. Question is, how far down the rabbit hole you want to go?

By the way, it is Infinite!

I still stand by what I said, the more you know about you – the less there is to know – full stop! But it is sort of a magic trick because there is no end as it's infinite. Best of luck on your inner journey! David!"

"Hi David, I've taken the time to read all of the pages you gave me a few times and I REALLY enjoyed it. Some of the things that I read, I couldn't even believe how accurate it was. For example it says that my ideal career path is nursing, medicine or psychiatry - which is crazy because these are the three things I am applying to in university. I am mind blown! If you would still like to show me the other 40% of your number magic, I would love to have a chat with you sometime to learn more about this stuff."

I replied, "Wow!! So happy! Thanks." So I sent more and more of the Science of Numbers and signed it off with. Pythagoras was indeed correct, when he based his teachings upon the philosophies that the "Universe existed upon Numbers."

I also said, "Oh yeah, remind me sometime whilst having a drink at Dodger's Bar, remind me to tell you about energy, about how to Manifest and how to manipulate energy. Once you learn this master trick. No one can ever take it away from you."

My student blows my mind, she doesn't even know who she is yet but let me put it this way, it's only possible to have master numbers on stages 3 and 4 and she has both. There are only primary numbers of 1 – 9. There's an infinite amount of numbers between 1 and 2. For example you have 1.1, 1.2 etc.

There is a bigger infinite amount of numbers between 1 and 3.
So, were discussing infinite which is infinite and now where talking about bigger infinities than other infinities, I can't explain it but it is so. Okay, Eveline should start 'mastery' starting at age 40 years of age. All the numbers are primary numbers ranging from 1 – 9 and then you have master numbers, 11, 22 and 33!

There are 12 types of people in the Science of Numbers. Each person has there own personality and people are far from just numbers but numbers play a huge importance in one's existence! Numbers can explain so much of a person's influences and it's infinite as there is no end to the 'rabbit hole'.

"Hi David, thanks for sending me more stuff! I'll get to read it after I've finished work. I will remind you about the energy next time I'm around there and you can explain it to me. Definitely interested in hearing about it. I'll let you know what I think about the new stuff you've sent me as well when I read it.
Thanks again!"

I replied, "Anytime! Always here in your phone! Ok . . . Here's one for you. I'm building a Bridge! And I want you to be the first one to cross over it. Now – you're about 10% over the bridge so well done! One of the steps needs to involve, not just energy, thoughts, vibrations and frequencies, but the illusion of Time! Once you grasp that concept, it's more than mind blowing, it's soul awakening but although it's a very big step on the bridge, it's almost like a bridge itself as once understood, there's no two ways about this, once understood – you will cross a bridge within, from mind to (you) the real you, your true self, the energy that feeds the mind, your consciousness. Once you connect with that energy, the real you, as time and mind are inseparable, once you experience the illusion of time, you will know that you found yourself and have woken up from a sleep walking experience, a dream that you once saw as reality.

So, I've experienced this already and wrote about it before. But for the over all experience, let's try getting you over that bridge. Who knows, it could be the best bridge you cross in your life! Cheers."

"Hi David, that sounds difficult to do but I'll give it my best shot. If I can do it though, I look forward to crossing that bridge. Have a good night!"

"Hi Eveline, Oh my God, you've just helped me as I've been up a lot of the night, making this bridge. I've surpassed myself, am beside myself as all of my previous 'out there' amazing experiences are coming back to my memory now and I'm blowing myself away. Not quite sure what this stuff will do for you but all good though, one step at a time.
You can be my first student!
P.s you're turning me into a teacher!"

"Hi David, I'm glad you're remembering all your amazing experiences. Look forward to hearing about some of them :)
I was planning on stopping by some time this week but I've caught the worst flu so it might be a bit delayed. Hopefully just a few more days and I'll be better and we can have a chat about all this number stuff!"

A few days later . . .

"Hi Eveline,

1) How did you go with the 'laws of life' make sense?

2) All the people only use 7 to 10 percent of their minds. It's a scientific fact. How would you like to learn how to use a lot more than 10%?

3) This is the journey I'm inviting you on. The choice is yours.
You have two choices in this game of life; you can sleepwalk through life and get 'played' or you can become totally conscious and become the 'player' that you were born for. This is my gift to you – the choice is yours. Okay, I've got my bridge made. There are 12 steps. The Science of Numbers was step 1, for step 2, what subject are you up for next, I mean, I'm on one side of the bridge now and you're on the other, so what's more appropriate for you, for step 2? I'll throw out a few options, Thoughts, Energy, Vibrations, Frequencies, Illusion of Time, Manifestation, The Now and the Power of it. The list goes on. Each subject is infinite but I'll just show you about 75 percent, (for us to understand) or so with each topic, as it's enough to cross the bridge."

"Hi David,

1) I read the laws of life and I really enjoyed it. It's easy to forget some things like that so reading it really reminded me on how to think and treat others.

2) I have indeed heard that fact that we only use 10% of our brain so I'd be stoked to learn how to use more, if possible.

3) I didn't know life worked that way (play or be played) so of course I would be happy to learn how to 'play' life rather than the other way around.

Well out of the subjects that you've shown me, the now and the power of it seems pretty fascinating to me so we can start there maybe?"

"Hi Eveline, Okay, that's a huge step though, I think energy has to be explained first, can you do me a favor, I need a name for this bridge. Perhaps we just come up with one if I successfully get you across it!"

"Hi David, Okay, well you can teach me what you think I need to know before we go on to the now and the power of it :)

Hmm, yeah I'll try to think of a name or we can come up with one when we're done since I'm not very creative and have no idea what to name this bridge."

"Hi Eveline, I must give you a heads up though as crossing this bridge, you might get crook a few times, nothing bad, it's just the energy changing within. The body might get colds, flu, run down but you come back to better health, more energy and so much happier and stronger as it's a change within. And if you don't go 'within' you go 'without'. And in your surroundings with people and relationships, once you get to a level of understanding, energy attracts and you attract this all around you and all sorts of stuff may happen, until things settle down again.

Life is a game but to understand, you have to see the over all picture, which is the rules, the influences, the understanding and the mechanics, okay? Also, I just want to get this out at the start, so we can recap at the end. This is a personal question! Can you sum up your beliefs about life and who or what controls life, before we start, as I want to re-show you your old beliefs once you're over on my side of the bridge. Does not matter if you think your beliefs are great or silly, it makes no odds, we recap at the end and we see the changes. This would be very interesting and this is the only personal question I'll ask you."

"Hi David, Okay, well I was raised in a Christian family so I do have religious beliefs however not so much for the reason that a lot of people believe in it. I'm not sure if there is a sort of higher power or not who controls things (although I like to think that we are beyond science in a way) but I do believe that having a religious belief can always remind me of how to treat people and how to be a good person because in the end, most of the rules that come with religion should be followed regardless (example: treat others nicely, don't steal etc.) So to answer your question, I haven't made a decision as to whether there is a 'higher power', as much as I'd like to believe that straight away but I am religious because it not only taught me how to be the best person I can be, but it's nice to think that there could be something after this life. Hope that sums it up for you :)"

"Hi Eveline, Perfect – great belief system!
But on this journey, every thought and belief in your mind, will be challenged. But this is good as you hold onto that what you want to hold onto and what beliefs that no longer serve you, you disregard and replace with some steps on this bridge. Lets get started."

Step 2

Energy

1) Probably learnt this in school. Energy can neither be created nor destroyed, can only change form.

2) Everything is energy, not just the wind, the oil, or cans of Monster but everything and everyone. It's impossible for anything to exist without energy.

3) Now, pick an item around you, look at it and answer this question. How many times can you cut that item in half before it ceases to exist?

First it's in half, so on so forth, then atoms, molecules, subatomic particles – once you go subatomic anything, all rules are out the window but keep separating in half and half again and tell me this, when does it cease to exist?

I know the answer but I tell you this, a belief will change in the next line, as there is no smallest, there is no biggest!

Your higher power, if you have one also has a higher power and so on, so forth. It's hard for our stop – start minds to comprehend this but it's infinite and I think you've just seen through to infinity. Well done!

Space itself expands at a rate we can't possibly measure, there is no stop point, where there's a little garden with a bearded man standing there, it's time for us to leave old beliefs behind (they worked in the last century but this is a new century where science is surpassing religion each and every day) and expand your mind. Infinity and expansion is what's happening.

4) Not only does rule 1 apply to all of what you see out there, what happens in the lab and what happens in nature etc. It also applies to people! And how to see this energy in people, broken down to its finest form, is numbers. So rule 1 also applies to you! You will always exist – it's a universal rule.

5) Do you know everything moves, everything is alive, people, animals, nature, even the ocean is alive and the earth known as Mother Earth is alive – how? Energy! Did you know – the stillest thing we have is a rock but inside the rock, the molecules are moving so fast and there we have the unmoved – mover!

Everything is alive as everything has energy.

6) The thoughts Step will show you how to shape, mold, attract and manipulate energy but you need to understand first, nothing exists in this world without either having energy within it or enveloped in energy, or both!

7) People have Auras. This is their energy on display for some gifted people to see. There is also a machine that was invented twenty years ago that can take a picture of your Aura. The colours of your energy change when you change but it's deeper than emotions that you feel. It's a person's energy!

8) Like energy attracts like energy. Opposites attract too.

9) Attraction – it depends on what level of energy you are on. There's no start and no end. It's infinite like all these steps. But if you're a life path 6 person – you should have a big family who you are close too, 6 is

all about family and falling in love and making a family, so a 6 person would attract that in their life from birth to death!

Now, if you're an 8 behind, inside I should say, the 6. Then the first 30 years of your life is about being independent, you'll try, but you might fall or stumble a few times but at the age of 29 and 30, you can look in the mirror and fix your crown and smile as you see success! This is energy and numbers help to explain the vibration you're on but we'll get to that later.

If you're going through issues, you attract people and places to help you over come those issues, you attract people on the same level of energy, you come together and work through whatever it is and usually go your separate ways afterwards, as this is how it works. There's no right and wrong, there's just what works and what does not work.

Like energy attracts and opposites attract too. For example, deep within you somewhere, you need to know this information and I'm totally full of this information from about 29 years of full on 'out there' experiences, so here and now, opposites are attracted.

The higher your energy goes, it works off vibrations, the greater the life you will have as you attract people on the same wave length and no negativity can enter the pure consciousness. And once your over this Bridge, Wow! – But one step at a time!

10) Your frame of mind greatly influences your level of energy. Ask and you shall receive might have been a teaching in your Christian beliefs and it's true but there's a catch and that's Energy. When you ask your higher power for something, you do so by being focused or saying aloud but if you have great emotion behind that request, your higher power listens to energy and it works, more about that in later Steps as we got to stay focused – Energy!

11) Energy is in each of us as we are energy. Now, you'll come across people who are 'Doer's' and others who are 'Talkers'.

The talkers don't really get around to 'doing' because by talking about it, they have released that energy and their minds move onto another dream or topic to talk about. The trick to 'doing' is to hold that idea in your mind, don't talk about it and release the energy, stay focused and get it done!

I'm just saying, the more you talk about doing something, the more energy is released and you're minimizing the chances of 'doing'.

12) Good vibes and bad vibes off people – pay attention to that as it's your deep self trying to get your mind to listen.

"Hi Eveline, I'll start working on the next step. Step 3! I think 'Thoughts' will be a good one for step 3. You see all of this information is in me, experience is my teacher and I'm just bursting with information at the moment. But the only way for me to do it properly is by Steps and only way for you to get over the Bridge is by Steps. So let me know how you go with Step 2, Energy. If there are any questions about any part, you just ask me and refer to the number it's part of, then we get through that and straight onto the next step. Very exciting step!"

"Hi David, Wow that energy thing was so interesting. No questions as I think I understood most of it after reading it a few times. Should we start on thoughts now or is there anything else about energy I should know?"

"Hi Eveline, I could write books about energy but I like to cut information to the core and just get on with it. So, if we're good with Step 2 Energy, Step 3, Thoughts is next. I'm blowing my mind away here preparing Step 3 for you because just recapping on previous experiences is quite good now, as over the years – the emotion has left me, I'm no longer affected by what happened and looking back upon previous experiences from a new improved light!

Let me some it up, I'm enjoying preparing step 3!

Thoughts are not just restricted to what I wrote – thoughts are probably the most important step!

To sum it up, thoughts are, after years and years of experiences, thoughts on a very deep level are in my opinion, when cut to the core, the only real thing that really exists – If that makes sense?

You might think – Okay, I have 60,000 thoughts each day in my mind, you might think, so what, but, if you ever wanted to find some sort of foundation to build up your mind, your belief system, not even to mention how your own energy exists in this world, to sum it up, your looking at Thoughts! This is the core of existence and were only about to tap into Step 3, thoughts! We still have approximately another 9 Steps to go."

"Okay Step 3 Thoughts is prepared. Now this is probably the most important Step and it's very important that you understand this Step before we move on to Step 4, Vibrations.

There may be talk of solutions to states of mind, that we'll cover in another Step and in later Steps there'll be talk of this Step too as Thoughts are really the very core off all existence, so enjoy."

Step 3

Thoughts

1) Each man and woman and child has about 50,000 to 60,000 thoughts in their heads each and every day!

2) Not only do your thoughts affect your emotions but they also play a huge part in your life.

3) Anxiety and Depression

Thoughts in the future create anxiety and thoughts in the past create depression.

Like energy thoughts attract like energy thoughts!

Once you keep making those thoughts in your head, 60,000 of them just waiting for your mind to grab hold of, after a while you see, your negative thoughts are attracting the same 'like energy', the same sort of vibration that the previous thought is on, you're attracting like energy thoughts and that's all you can feel and experience. Once you keep thinking this way, you end up in a spiral, like energy thoughts attracting like energy thoughts, getting greater and more powerful and if you don't get out of that programed spiral, self harm and release unfortunately becomes the only option as thoughts have a huge affect on the body.

The thinking needs to be changed, or even better still, becoming present which is a later step.

When in a frame of mind, unnatural, unbalanced I should say, not just your mind but also your world becomes this depression as whether you talk or not you attract people who only talk about depressed stories. You help them remember the time they were sad and depressed as it's all the same energy – the same vibration that you're on and all that comes into your world is more sad stories, more depressed thoughts and more negativity to put in your mind, your thoughts and your energy.

Anxiety is your mind thinking of the future, it could be imagining work the next day, it could be a dream so far fetched that your trying to live in your mind and exist in the present at same time, that stuff tears you apart as your trying to be in two places at once.

These frames of mind (some people say the person has a week mind but that's not true) can kill the strongest of men! No kidding, this stuff is not a joke. But I see – I see what's happening.

4) Thoughts create the emotion in your body. Your emotion is energy in motion basically. Your thoughts affect that greatly. There are other feelings that come from inspiration and that's beyond thought. But yes, thoughts on top of thoughts and so on so forth can create some very unnatural emotions. Think good thoughts, when thinking good thoughts, it affects your whole appearance, some people may still have a crooked nose but their energy, their Aura makes their appearance so attractive and radiant. What you think, you become, so mind your mind.

Always search for inspiration, it's out there, it's within, once connected, once your thoughts slow down as there's a place in between one thought to the next. In that place is so much happiness; it's so great and it feels so good, as it is your true self, your consciousness, your energy and the real you, unaffected by thought and what thoughts do to your emotions. The mind can only live in the future or the past or solving problems or learning or conflict. It cannot exist right now, without thoughts of future or past, which creates anxiety and depression. The Now Step could be the best step to talk about, in fact, I'm sure it is, it's well beyond all the other steps but one step at a time. You'll get there!

5) Your thoughts are creating your reality on so many levels, not just thinking of the future and creating anxiety within, not just living in your past and slipping in and out of depression (making the prescription medical world a fortune).

But your thoughts are actually creating your reality. The present moment is the moment where you pre-sent your thoughts out and receive that as your reality.

Most people don't have their own thoughts, they just receive others and pass on other peoples information and call other people's theories theirs. Thoughts come into the mind and they leave. If you entertain it, by giving it energy, it will be manifested in your mind and soon your reality. The question of all time is where does the thought come from? The answer is thoughts are just getting past around to each person by interaction, media, internet and TV but creative thought is coming from the inspiration from your higher or deeper self, your consciousness and the essence of life. Your mind receives the inspiration and creates a thought. Create your own thoughts from your own consciousness and create a great life for yourself.

You might see (believe) your higher power is answering your thoughts and that's a Yes and a No, as it's you that's sending the thoughts out there but you are part of the higher power.

Imagine a force out there, yes a different level of energy vibrating at a speed so fast, it's just so intelligent, it's beyond us at present but this force gives you multiple copies of your thoughts that you send out, usually the more emotion behind the thoughts seems to be quite beneficial, so you send this level of energy out, thoughts too are energy and up it goes. The force matches that and gives you a copy. Now this could be anything in our world to anything as it's infinite. But you attract people, places, situations, hurdles to over come and fears to face.

Fear attracts like energy fear, not just any fear but the exact fear you fear, as it's an intelligent force, it gives you multiple copies of your thoughts. Now let me tell you this, don't think of that fear and you won't attract it or second choice, think of it, attract it and over come it, facing a fear can be most rewarding.

6) Justifying and Choices. The mind – each one of our minds justifies each and every action we do and not just that but our minds justifies each and every thought we have in our minds, no body is doing anything wrong given their model of the world, their experiences and what life showed them. Makes you think hey!

With Choices, we choose before hand who we're going to be that day or evening. You can sleep walk through life or you can choose. The trick to mastery is to keep choosing who you really wish to be and in your mind choose and choose again, same thought each time, who you would like to be and you become that person. Sort of like 'fake it till you make it', but by choosing.

7) Seed planting, in Australia back in the 80's a girl was hired who made effective advertisements. Her job was to make an effective advertisement for Holden who sold cars. She was able to make an ad that planted a seed into the minds of most young Australians. It worked perfectly and in twenty years time, so many young Australians were buying Holden cars and to this day, Holden is the number one car in Australia. Once a seed is planted in a persons mind, it has to grow and come to fruition eventually. Planting seeds in the minds of young children is very effective and many companies do it in today's world.

If you plant a seed in your mind of success or failure or what ever it may be, plant the right seed in your mind, think about your goal, see yourself achieving your goal and you must succeed. Every one of us is the sum total of our thoughts as we are guided by our minds.

The mind is like the land – it doesn't care what you plant, success or failure, a concrete worth while goal or confusion, fear, anxiety and so on. But what you plant, it must return to us. As you sow – so shall you reap.

8) What you think about expands. If you hold onto a thought, whether depressed or worry or insecurity or what ever it may be, the more that you hold onto it, the more you feed it energy, the more it will expand in your world and you will convince yourself that you have serious issues. The truth is, it's just a thought, expanded.

What you 'Want' you push away, for the very act of wanting, you receive that – the feeling of wanting. You need to accept you already 'have' and you attract. Once you believe you have it, could start off with money and more of that gets attracted to you. If you slip into debt you know you owe money and you attract more of that, so hold onto at least 10% of your money, know and believe you have money and you attract more – too easy!

You can train a frame of mind to think anyway you like. It all starts off with a thought and perhaps in three weeks your mind has not only justified each and every thought, it holds together approximately 60,000 thoughts of that frame of mind per day. If you sow negative thoughts, your life will be filled with negative things. If you sow positive thoughts, your life will be cheerful and positive.

You are what you think you are! You simply cannot kill thought, we simply change form and we always exist, as we are intelligent beings full of thoughts.

9) The brain cannot produce a thought on its own; it's a scientific fact. The brain is the mind manifested as material form. This means if you produce a thought, you have a mind and real essence in you. We only use 10% of our brains as science has proved. I personally believe that the more essence we acknowledge as us, the more aware that we become of our consciousness, the more activity our brain will be doing when we choose to use it.

The Father, the Son and the Holy Spirit are your Mind, your Experience and your Energy. Once your energy gets inspired, you produce a thought in the (father) mind and then you feel the experience (the son).

You should focus on your dreams – don't tie it to a person but to a goal, keep that one thought in your mind and remember this conversation, when the goal becomes your reality, you'll see, thoughts create reality too!

10) There is a collective consciousness out there. Approximately one percent holds most of the wealth as they have broken away from the masses out there and learned how to be successful, although this success is passed on from family or cooperation through generations but it's kept going. Conformity is people acting like everyone else and not knowing where he or she are going, but just sticking with the collective consciousness like everyone else.

We live in a system created universe and we receive multiple copies of our thoughts. The collective consciousness also receives multiple copies of its collective consciousness, from towns to countries to the world.

Some people don't believe that your thoughts create your reality. That is because many are just thinking with the collective and assessing their memories of the past and living in their minds projecting the future. None of this is creating your reality. It is not just your mind (as in your head) that manifests. It's true, but it's not fully explained, there's a missing link.

For creation, you need to clear the mind of future and past thinking, to search for a moment and to produce a thought on a deeper level but really it is just avoiding the constant thinking that we all do and placing all that wasted energy into a new idea, a new concept, a new thought on something that we wish to draw into our lives. If you do this from the heart or from the spirit, there will be no problem and if you do it from the mind, you need to put almost all of your energy into it to make it clear, to really focus, to visualize. It's the emotion behind the thought that fuels creation but more about this later.

11) If you're ever asked the question, what really exists and what does not – you can answer – nothing exists but thoughts! Basically, what you need to know is, we become what we think about. A person is what they think about all day long.

A person's life is what their thoughts make of it. All things are possible to a person who believes in it. If you think in negative terms, you will get negative results and if you think in positive terms, you will achieve positive results.

You will become what you think about. Plant a seed in your mind of success or failure, your mind will grow that seed whether you think it's good or bad. Plant the right seed in your mind and think about your goal, see yourself achieving your goal with your minds eye and you must succeed. Use thoughts and imagination to create a goal and follow that goal. People with goals succeed because they know where they are going.

Every one of us is the sum total of our thoughts and were guided by our minds. We must control our thinking as we can use it for good or bad. Life should be an exciting adventure but as you believe, so it will happen to you. Remember the word imagination and let your mind begin to soar, concentrate on your goal each day and use action, as ideas are worthless unless we act on them.

As you sow, so shall you reap, which means what ever you put in your mind, that seed gets planted and soon it will be your reality. All you need to do is have your goal and know where you are going; the answers will come to you along the way. All you need is purpose and faith. Don't worry as worry brings fear and fear is crippling. One of the greatest discoveries is that people can alter their lives by altering their attitude of mind. Success is the progressive realization of a worldly ideal.

Now if section 12 does not blow your mind away you must get back to the start and re-read.

12) 60,000 thoughts in your head, means that you are dispersing your energy, which is the real you, (covered in step 2) into 60,000 places each and every day!

Here's a trick, a magic trick, do not ever look upon this as failure if you can't do it, the other steps will show you how but I'm just planting a seed in your mind now on Step 3 (all seeds planted, grow).

If you can turn those 60,000 thoughts, into One Powerful Thought, you can manipulate energy right before your very eyes. Remember everything is energy, like I said, the other Steps should show you the way!

"Hi Eveline, you're going to like step 4 Vibrations and then Step 5 Frequencies. Step 6 is not invented yet but I need to present to you more or less the order life presented to me, so let me know how you go with Step 3 thoughts and if any question on any section, write the number beside the question and we will discuss. I just want you to be happy that you understand and after that we move onto Step 4. All steps are exciting and all steps are infinite!"

"Hey David, I finally got to read all that stuff. So firstly I found Step 3 really interesting about the negative thinking spiral and how it keeps going unless you change the way you think. I found this so great because I studied this at school about a week ago and read a study about negative thinking on depression and anxiety and it's actually proven!

Secondly, step 12 did blow me away! I have no idea how to get 60,000 thoughts into one thought, which projects heaps of energy. I understand the concept, just can't do it yet but we will try and get there.

I think I understood all of it so no questions for now but I did have to read it a few times to wrap my head around it. Ready for the next step I think when you are."

"Hi Eveline, I'm so proud of you for been able to handle this stuff. With regards to the 60,000 thoughts brought together as one incredible powerful thought – we should revisit that on the last step but I like planting seeds, so I'm working with your mind. All steps need to be taken first before real Manifestation occurs!

Once over this bridge, you'll have the tools to, let me just say – a whole new world of unlimited possibilities!"

Eveline and I met up for a chat and she was saying that the whole idea of using 60,000 thoughts as one great thought in your mind sounds great but feels it's impossible. So I said, let's do an experiment – just between us two and using your mind as a tool on Step 3.

So between us, we ended up planning a night out. Casino does a free movie on Monday night and the restaurant does seafood for $10. So I said, "Keep thinking that you, I, my partner, my kid, your three friends go to this Casino on Monday night. With the next few thoughts in your mind, keep thinking the casino with your friends and my family and I. Put energy into that thought and make it expand. Then once you can visualize that in your mind – pay attention to detail, imagine the food, even the smell of cooked seafood. Then even try imagining the movie, what movie could be on that night. So on, so forth, the more detail the better. And spend quite a lot of time thinking about this night, the first night that were all out together! But continue on with your life and get all done that you need done but come back to visualizing this night every so often until it's in your life. I won't say a word and you shouldn't either, test this mind of yours and sit back and watch what you bring into your life, you might just shape energy and attract that very night with the right people into your life. Everything is energy."

A few days later . . .

"Hi Eveline, as we spoke the other day about thinking the same outcome in your mind over and over again, you won't have to think it with 60,000 thoughts, even 100 max a day would be well more than enough and it's easier to just focus on the night. And the more detail the better, for example try imaging smelling the seafood at the restaurant that night and perhaps what movie could be on. But over all, the main trick to manifesting – is visualization! So try visualizing the night in your mind, the people that are there and picture something in your mind, or an imagined smell and don't tell me and we'll see what happens! Your world should change if this works for you, that's all the tips I can give, so you're on your own now in the manipulating energy world by your mind (the tool). The future will tell!"

Eveline kept thinking of the Casino and soon enough her father asked her did she want to go and get lunch there for the first time ever.

Her friends started talking about going to the Casino for a night out, as they have never been before. I however got so involved in this book, that those around me, decided to do a sort of intervention on me and take me to the Casino to try bring me back to the previous years where we used to watch movies and have dinner there. So the Casino was brought into our realities just by spending some of the day focusing on it. Eveline was amazed, particularly by her father and the lunch they had at the Casino.

Step 4

Vibrations

1) All that really exists is Energy – Thoughts – Vibrations and Frequencies.

If you some these up further – you're looking and an intelligent form, of vibrational energy that uses frequencies as tools.

2) The faster your energy vibrates, the more aware you become. There is so much of you there and so much you're not aware of.

Life is not only an adventure; life is about experience and self-discovery.

The more you know about you – the less there is to know full stop!

3) I first discovered Vibrations when I was about fifteen years old, I was constantly in my mind trying to find one core thing that is a sort of building block for all of life, something that connects everything and everyone, something that is real behind the illusion, deeper and deeper and deeper I went and I couldn't seem to find a core connection until one day it just came to me. I think I was about to give up searching and once you stop looking, it just comes to you. The radio was blearing out great music and I could feel the vibrations off it, I could almost see the vibrations in the sound as I was so deep in thought and vibrations

became the answer I was searching for. Music, trains, engines, electrical storms, rocks under a microscope, places, people, everything is vibrating at different levels, faster or slower and within each person is an over all vibration and numbers can match each vibration. Numbers are infinite and also are vibrations!

4) Vibrations are energy and are calculated by numbers.
The levels of energies are broken down to the vibrations each energy is vibrating at. For example, depression is a level of energy that ends up slowing down its vibration by the use of thoughts and when you're out socializing or just scored a goal or for some people church music and for me, being totally inspired and totally present, our energies are vibrating so fast, it's so enjoyable, were so happy and we have found our true selves! But even that last sentence is infinite as there is always more, always more self awareness and I've got so far, I'm sure it would blow your mind but even I only got so far, I see you in this life time getting beyond that, okay, one step at a time!

5) Trust in yourself that you will change, higher your vibrations automatically, as you're in synchronization with life, your consciousness if you will, always is in total tune with life, you just need to connect every once in a while as it's just the mind that spins out of control and imagines the worst case scenario at times but your consciousness knows, the more present you are, the more in-touch you are with your true self! So, no need for anxiety, trust in the game of life, I'll give you all the rules and insights and when the time comes, you'll be in your element! The exact vibration that you need to tackle, the same event or situation that you brought into your life by using your mind as a tool, you'll be prepared for.

6) A close family, even a father, daughter relationship is one of the best vibrations I have ever been on as people can come into your home and deep within they just sense that vibration of energy – it's inviting on all accounts! Each of us is emanating a vibration, which is the result of our predominant thoughts and beliefs in each given moment. When one person is giving off a happy positive vibe and the other a negative vibe, when they come together they will affect each other's vibration. The positive will slow down to the negative vibration or the negative

will speed up to the positive vibration or the two people will not be able to stay in the same space as their vibrations are literally repelling each other.

7) There is a universal law that everything vibrates. Everything around you is vibrating at one frequency or another and so you are too. This includes your thoughts that are made of energy vibrating at a different frequency and the same with your emotions, which are energy in motion that vibrates at different frequencies. Once you know that you are energy, same as your thoughts and emotions and you understand that whatever vibration you are sending out there, gets attracted to you as energy attracts and it's the different levels of vibrations that get attracted to you, matching your vibration.

8) Once you're on a higher vibration, you get to work out so many things, all the clouds leave your mind and everything becomes so clear and you get to see why each experience happened in your past, everything comes together, everything makes perfect sense. All that is in your surroundings is also in synchronization with your vibration, which is confirmation on all that you realized, become aware of and now know to be true.

9) A trick to manifesting your reality is to use your imagination, hold onto a dream and become that dream, as you are what you think about most all day long. Once you hold that dream in your mind, you're sending out thoughts that are vibrating on a level of energy which is also affecting your feelings and that attracts the same vibration that you are sending out and you manifest your own reality.

10) Once you're in The Now, a later step, your energy does not get dished out to the 60,000 thoughts a day that it does now, you become present, all your energy is present and your vibrations are so fast and aware, a pin could drop in the hallway outside your apartment and you hear it. When your energy is not dispersed into 60,000 places, you're totally present and you know everything in your environment and you see how energy is been shaped and influenced, you rise above this mind illusion world and you see, hear and sense everything. No drug can do this for you – this is on the part to enlightenment and all starts off with

higher vibrations and were back to, the more you know about you, the less there is to know – full stop!

11) When your vibrations are higher, you affect most of the people around you as they feel this vibration and can't explain it, without evening seeing you. It's on a deep level but they feel so happy and excited and depends how higher, how faster your vibrations are going, they could even seem so 'out there' but so present, it's like looking at your own energy in the mirror. It's a total wake up when you have moved someone's energy to your level and then you know this is all so real!

Some people don't get affected when you walk in and your vibrations are very high as those people are so identified in their minds, to them, they are their minds but others go wild and they say things to you that are so out of character but totally present. You have brought them to high vibrations and they know it's you. Usually this works better with females as they are always half in-tune with themselves, their energy, where as some blokes can be more lost in their minds.

Also animals have a 'knowing' when someone has a good heart though their vibrations and this makes them feel safe being close to that human.

12) Always try to higher your vibrations – some times you might raise the vibration too fast, too quickly and then your feeling flat for a few days. It's a bit like the wave of an ocean, if we see a really high wave, a deep low trough balances it but every mood and feeling is temporary until you get to the Now but in the meantime, remember this – This too Shall Pass. T.T.S.P.

"Hi Eveline, there is so much in the Vibrations step that many books could be written about this step. This is infinite like all other steps. When a person vibrates at a fast speed, that vibration affects all around them, all material objects are affected as each and everything is energy, it's all just vibrating at different speeds. There is no good and bad vibration, for one makes the other exist, all is, what it is! Ok – if there are any questions about this step – write down the number it's referred too and ask away. If all is clear and you feel, you don't need to know anymore – we can then move onto Step 5 Frequencies, very interesting step as it's tied in with the sixth sense, your intuition taken to a new level."

"No questions at all, everything sounds good. Eveline."

The vibrations have been raised in my house without people really knowing what's going on but Eveline is pushing teacher and I'm teaching my student some good material, so the outcome between the two of us, is higher vibrations within and our environment and every one is getting colds, flus and coughs! But I see, as I've done all this before ten years ago – I see we all come back happier and healthier and on a new vibration. Once more aware and once a higher vibration, no one can take that away from you!

Step 5

Frequencies

1) Frequencies really exist – it is not something that's an in-figment of someone's imagination. It is not a belief – it is very real.

2) Lets start off small, when listening to the radio, the station you are listening to, the sound you hear, travels on a Frequency. You could be listening to 105.7 on FM in Queensland, that's a good station at times, less ads and more music. The frequency is labeled by a number – just like vibrations! FM radio has how many frequencies? Yes you're catching on – it's infinite. Then we have AM radio and that too has many stations that we don't really listen to. But it's the same too, it holds frequencies that one person can sing into, using technology and that voice which is energy gets travelled through this frequency that consists of waves and energy and with technology on the other end – you hear great music coming out of the radio. Very straightforward really! Now how many frequencies are on AM radio? Yes – correct – it's infinite! Well done!

3) Now – do you think that people are somehow above the universal laws? Some people seem to think that they are – where are their heads? I think that those people get so involved in frequencies that they see it as magic that only affects what they see and hear but failure to look at the Self too. So here we have it – people too are on a frequency! As we are on FM frequency for example, there are also other frequencies too.

So I ask the question as we are on FM frequency, how many frequencies do you think is really happening in our world at one time, in the one moment? Yes, why would frequencies be restricted to just one or two – you seeing the bigger picture here? I personally know of three different frequencies but I see like all Steps – Infinity!

4) Frequencies work for a living.
Frequencies are not only one of the main building blocks of the Game of Life and not only the Intelligence Vibrational Energies tool but Frequencies are always working and always existing, not only infinite but also pure genius!

5) Frequencies travel perhaps just as quickly as the speed of light! Frequencies don't have passports – frequencies are not just a tool – frequencies rule the traveling energy world!

6) I love frequencies and frequencies allow me to sometimes get in touch with AM station. And I have seen gifted people in action doing all sorts of stuff by the use of frequencies!

7) People (intelligent vibrational energy) with bodies are on FM and other people (still intelligent vibrational energy) without bodies are on AM.

Gifted people on FM can talk and hear people on AM. It's truly amazing to see and to experience this.

Now – the best way to explain this is I'm a 7 vibration and being a 7 vibration lets you become aware of, one could say, invisible forces that are out there and 7's can feel this be means of what we call intuition or the sixth sense. Everyone is intelligent vibrational energy and everyone has the sixth sense but not all people use it. I'm guessing 70% of people don't know or are aware that they have a sixth sense! But I use my sixth sense and I gather information as that sixth sense is connected to an invisible force called Intelligent Vibrational Energy! So, I feel and know the information by a way we call Inspiration and that rises in me instantly and my mind gets the information and understands it. Then I see, I know stuff that the minds of the world can't work out. I couldn't work out either using my mind but the sixth sense awareness bypasses all sorts of mind rules, laws and illusions! This is still frequency stuff, as the sixth sense feels and also see's frequencies such as AM and I'm going to call the other one I saw (from within) AAM.

8) So there's an infinite amount of Frequencies but I only know of FM, AM and what I call AAM.

AAM, I'll explain in Step 7, when life showed me, so there's no confusion as how to see frequency AAM. Totally and utterly an amazing frequency that it vibrates so fast, there is no form of negativity that exist there. I always felt that negativity and positivity must exist in the law of Life because one makes the other exist but AAM is a totally different frequency that seems to have different laws as negativity doesn't exist, some people feel this frequency, like I did, as pure love!

9) FM and AM. Thought is energy, as you're on FM frequency, you think of a person who passed onto AM frequency and they're attracted to that thought, they become where you are, so be aware of what thoughts you send out!

Energy and thoughts we covered in step 2&3. But having a sensitive nature and lots of 7's in the Science of Numbers, one can feel energy from someone who has passed to AM frequency, it's a gift but numbers explain that gift. There are people who can use one of their senses, which is very sensitive, example, some people can see others on AM, some can hear others on AM, trust me, I have witnessed this, some can feel energy on AM, which is what I could do and could not switch it off for approximately twenty-two years. During that period, by the moment I was on Step 7 – I had more friends on AM frequency than I did on FM frequency! Not once did I get a bad vibe from any energy who was once a person, all connections and feelings were totally amazing and I knew who each energy was by the way they vibrated, I could feel it.

10) When we go down on a sub-atomic level, we do not find matter but pure energy and all of it is vibrating at one frequency or another. Everything has its own vibrational energy, including a rock, which is the

stillest thing that we have. Also our thoughts and emotions all vibrate at a different frequency.

11) Frequencies are everywhere, they exist, just because it's an invisible force like gravity, does not mean that it doesn't exist. Sure, some minds of the world have difficulty accepting that we too are on a frequency but there are other ways to see this, to feel this by becoming present and aware. If you become your energy, your vibrational energy, if you become your true self which is Intelligence of Vibrational Energy, you can then feel, see and know of Frequencies. Until then – just tune into a Radio Station and listen to music and enjoy what frequencies are giving you in each moment. It's all infinite and Frequencies the tool and Frequencies the energy, gives us an infinite world of gifts in each moment!

12) If we are on a Frequency, which we are and if our mates who passed on are on another Frequency, which they are and AAM frequency exists, which it does as I have my experience and also experiments that I conducted and if there are an infinite amount of Frequencies, which there are – just play with your radio to see that. And if Time is an Illusion, which it is but must step out of the mind and in to true self to see this – explained in later Step. So if all is existing Now in this moment which it is and all is infinite – which it is and time, past, present and future from our minds perspective is happening Now, we are on a Frequency and we can move to another Frequency and we always exist, proven in Step 2 – Energy, so we can be in the present in a timeless reality and on any Frequency we like in the overall spectrum of existence.

As you start to go through the Steps, you get educated, as you become more aware, as you find out more about yourself, you start to see, the gold was always within you – you just forgot who you were!

All of this information is golden, is Infinite and once you become it, no one can take it away from you – you become complete and the whole world is your oyster!

Eveline "Can you change the frequency your on or is that linked to thoughts and whether you are positive or negative?"

David "Once you die you change frequency but you can also tap into other frequencies by raising your consciousness, your awareness that's explained in later steps."

Eveline "Ah okay."

Step 6

Science and the Universe

1) Space expands, how does that affect our solar system, when space expands, what happens?

Well, all planets are in their position so the planets move away from each other. If you measure the distance from earth to the moon – what the hell, venture out there, earth to Jupiter, the distance should be that much greater than that of a previous year! But Time is an Illusion, which I'll explain, in another step. So, what are we left with? Expansion and Energy and Everything in Existence is alive! Now, you ever heard "ignorance is bliss!" Someone who understood too much would have said that and although I salute that person, we must soldier on!

Let me tell you some of the stuff scientists are saying these days.

I've had a good chat with a few, I'm writing the bridge between science and our old and new beliefs but I didn't share much as I only questioned and this is the outcome!

2) They are thinking along the lines of this.

"We are not exactly sure what this fuzzy line between consciousness and unconsciousness and sleeping and dreaming, it's all going on in one cycle but we're not exactly sure what part of it we are playing all the time as we have to hit on things to be sure."

"How many times have you had dreams – oh it's a dream – how do you know that all of it is not a dream?"

"The whole idea you can dream, the idea that you can imagine, like, some mystic part of your brain moves things around and lets you have this cartoon dream, what the hell is that?"

3) "We are brilliant in comparison to cave men don't get me wrong but we don't know much as we don't understand the universe."

"Once we go to Subatomic Particles, all rules go out the window when you talk Subatomic anything."

"A particle can be in a Super State, where it's moving and is still at the same time, what does that even mean?"

4) "And the idea of if your looking at something and you change the behavior of that something, the observer actually changes the very atoms that are moving, how is this possible, we don't know!"

This is called the Double-slit experiment. The observer changes the behavior. It proves that measurement is everything. At the quantum level, reality does not exist unless you are looking at it.

"We don't even know why the universe stays together, why it doesn't just fly apart."

"I am thinking that this whole thing might be someone's imagination, it might be yours, it might be mine, it might be a combination of everyone's imagination. Things are real as there are real laws to this life."

"I am not sure what this is, I am not sure you do either, you might pretend you know what this is because it helps you sleep better at night. You can pretend that everything you can't hit with a hammer is not real, that it's just your imagination."

"What is consciousness? What is life? Everything on the Internet is broken down to 1's and 0's and then that turns into a picture or a video for another person's mind to interpret, 1's and 0's we are talking here and that's the core of the World Wide Web."

Then there are 1's and 0's within the fabric of quantum physics, the very essence that makes up the whole universe - is life a great big game?

"No one knows what electricity is or how the World Wide Web works, it's just there, existing."

"We don't even know if this is the only universe existing at once, it could be all existing at once and just our human minds see this on this frequency."

5) Scientist are now talking about, "Inside each galaxy is a massive black hole and inside each black hole is another universe. The universe holds together 100's of Billions of galaxy's, each galaxy with a super black hole and within each black hole is another universe with 100's of Billions of galaxy's, each with a super black hole and . . ." I say Infinity! - Continued, "we have no idea!"

Hold on now, don't get lost in the madness, scientists have got an idea. Look at this. "Imagination is the key – all is first created by thought."

6) "The universe could be infinite and each universe could have its own laws, we just don't know, we as in, not just you and me but we scientists in this Now of 2015."

I love scientists and science and I can see the more a scientist explores and learns, which he must as it's his job but the more he knows, the more he realizes the less he knows in the bigger infinite picture and Pandora's box gets opened and revisited everyday.

7) "We don't take our imagination seriously – we get a thought from the either or our imagination and that thought gets manifested into our world." Scientists are starting to believe that not only imagination and thought manifest in our world but that our own thoughts are affecting stuff in our world.

This is a good one, "We are Antennas, once we pick up on something, we see more stuff but it's all created by the human mind."

"We are made up of 99percent space, one percent of every atom is composed of protons, neutrons and electrons. The other 99percent is empty space."

To put this in perspective, imagine an atom the size of a football stadium. Place an orange in the center to represent the approximate size of the nucleus. You would then place pinheads around the stadium to act as electrons. These are the building blocks of our world and yet they contain very little substance. In fact if the space was eliminated and the matter condensed the entire human race would fit in a sugar cube.

8) I'm here to bridge that gap between science and our old and new beliefs.

Okay, enough said by others, I'm taking back over this Bridge.

This is the really real world, whether real or not real, this is our world and our best scientists are all over this but the further they go down this rabbit hole, like I said before – ignorance is bliss. This stuff is really happening in what you call reality. Your mind has an effect on what goes on in this world through thoughts and collective consciousness, which changes our realities.

Beliefs, for example the Placebo Effect, give a person a piece of sugar from a doctor, someone they believe in and get that specialist to tell the person, this will cure you and they get cured as beliefs are so strong they can heal you totally!

Change your beliefs and you change your whole world and as you are the drop in the ocean, that ripple continues! It's all infinity, I know this and I totally know this but still my mind gets blown away each hour of each day whilst writing this stuff. Once you even mention infinity to a Stop-Start mind – you're calling forth the energy, the consciousness that feeds the mind to tackle this.

9) The Universe.

The Universe is like a giant Xerox machine as it gives us multiple copies of our thoughts, whether positive or negative. Everything in the Universe is all part of one ocean of energy, all vibrating at different speeds and our thoughts that we send out attract that vibration to us. We swim in a sea of consciousness like a fish swims in water.

10) I have observed this all of my life, that my thoughts fuelled by the energy in me was bringing forth circumstances in my life. At stages I couldn't seem to bring anything good into my life as I felt that I was full of issues, negativity, frustration and all I could think of was, that I needed to end up in a fight or to get beaten up in order to balance out how I felt inside. Most of the circumstances that I brought into my life were on the same vibration as to how I was feeling and they were not pleasant experiences. To me it was so obvious as to what was going on, so it occurred to me that I needed to change my thinking. I was living in Ireland at the time and I changed my thinking to how life would be in Australia with blue skies and hot days and the feeling of being out doors most of the time, to me, felt great. This constant thinking of Australia, changed the negative emotion within me to a positive vibe over time and once I was finally on that positive vibe, I was then attracting people in that traveling field, for example, this girl came into my life and told me that she gets free flights as she works for the airline and over time, she gave me 1st class tickets to Australia for a very cheap price. As I was using most of my minds thoughts, focused on the vision of Australia, the more detail the better, I was attracting great vibes in my life. But my experience in Australia was much better than what my mind could dream up of.

Each time that I analyzed the situation that I was in, I would think about what I needed in my life, to feel happier, free, relaxed and at peace with myself and then I would hold that situation in my mind until I could visualize it. That's when the power started, new people came into my life, things changed and soon enough I was in that reality that my mind was thinking of and visualizing.

11) In order to manifest your reality, you need to slow down, relax and be present. Once you move away from the chattering mind and find

stillness within, you can then produce a thought of your own. Once you envision what it is that you would like in your life, you hold onto that frame of mind, which changes your vibration, your energy in motion and as soon as you can visualize what it is you need in your life, you have already manifested it and then trust in the unfolding process. Soon that event will be in your reality.

12) The Universe does not care whether your desired outcome is of a positive or negative nature, it sees it as your point of attraction and matches it, it gives you a copy of your thought process.

Eveline "I'm writing the bridge between science and our old and new beliefs. I think this is a really great statement and if you're going to get the book published, consider putting this on the back or front because I found that really interesting."

David "Thank you so much. I'll do that then if you think it's good. Cheers."

Eveline "I think its genius. If I were to read that on the front cover, it would get my attention. Put it in bold.

Next step?"

Step 7

Mystical Experience

1) Ask and you shall receive! Worked for me, once a massive amount of emotion was behind it, I was almost insane with frustration shouting at the sky due to some social and depression problems in Ireland and after that moment, my life changed totally! Soon enough I was on a plane going to Australia and once I arrived, I had the most amazing mystical experience of my life, it was surreal.

2) This Step for me was about releasing anger, stress and negativity.

It was about an inward awakening, about expansion, about transcendence, about being connected to the Intelligence Vibrational Energy out there.

3) An incredible surge of high energy came to me in waves and I became so alive, so aware and entered into a reality called a mystical experience.

4) All I thought about myself and my life and who I was since childhood, was just a blimp on the radar, my ego collapsed and my soul was set free for the first time. Everything that I thought of myself was just a mind illusion. I woke up and came home to myself. I had the feeling of a sense of oneness with everything. You feel that you are everything and everything is you as your consciousness becomes at one with everything, a sense of all knowingness became present.

5) All senses became heightened, improved taste, smell and hearing.

I'm a very intuitive person, life gave me a gift called a sixth sense, we all have this sense and for most people if they loose a sense like hearing for example they develop their intuition, their sixth sense. This sense became very strong within me; I could feel everything around me and knew everything from within. Also you can feel other people's vibes and know what they are thinking in a way like never before as you become awakened.

6) I moved into an awareness of timelessness, there was no time, everything was happening now and that's all that existed, a totally timeless reality.

7) Everything was sacred, everything around me became so sacred and powerful and crystal clear and alive. Most people use 7-10% of their brainpower, but at these moments, I have looked into this and found out that I was using 70% of my brainpower; I became all knowing as my spirit awoke within, after my ego collapsed.

8) My consciousness broadened to a global consciousness, a collective consciousness. Global issues that were once in the back of my mind before were right there in front of me as a suppressing issue.

As we talked about Thoughts in Step 3, let me tell you what I see in this mystical experience. I see a collective consciousness, not only does an individuals thoughts affect and attract energy and situations and people but I see that this collective consciousness is the global instigator for making, shaping, molding the world at present, not just global warming by expanding our minds which is affecting the worlds vibration but by changing our vibrations, causing natural disasters. I see all around me, the news, the changes that Mother Earth is going through is all to do with 'Thought' and this thought, collective thought is energy that is vibrating faster and faster the more we become aware of ourselves by expanding our minds.

9) I was downloading information beyond my own comprehension. Between energy, thoughts and frequencies, I could see another world

that was tied in with this world, just a different frequency, not just one other world but also two other worlds as well as the one I was in and all this was happening on this Bridge. I was in FM (spirits with bodies) and tapping into AM (spirits without bodies) and Arch Angel Michael's world all at the same time. That was an experience and a half. I could see three worlds, three different frequencies.

Then my consciousness showed me a future, showed me about Time the illusion, about so many other illusions that our minds and beliefs make real.

My consciousness showed me a language that I could see myself talking but never heard of before. (Later to meet a man who told me it's called Tongues - a Christian language.)

My consciousness showed me a future where I leave my body, where I see my Granduncle and see my Grandmother, life showed me so much amazing information, this field of consciousness out there was coming into me but more so, I was becoming aware of so much of my own self and the energy around me.

This awareness was affecting my mind and my minds thoughts were going wild with visions, future happenings and I was also trying to make sense of what is happening.

But all of this was happening so quick, I felt that I was about to burst into infinity; it was such an amazing, almost orgasmic feeling all over the whole body and within. My mind came to the conclusion, that I die and leave my body and I get to see my older family that too have passed on. Sure what else could you think?

10) Angels frequency AAM!
In this Mystical experience, there's an Angel world out there. Everything exists at the same time. Angels are energy, on a higher vibration, vibrating faster and on a different frequency.

I meet a person called Liz McDonald and to me, she is an Angel manifested in our world, in our frequency. She informed me that the leader of the Angels is Michael Arch Angel. Michael AA for short. So there's like a pecking order in this Angel world but it's not really like that, it works off energy and vibrations like all of life and Mike AA is more aware of himself, more aware of his own energy. His energy vibrates faster than all the other Angels, who all vibrate faster than us. So it's brings us back again to 'The More you know about you – the less there is to know – full stop'

If you ever want to get into this subject, fell free to contact Liz at *www. angelreading.com*

11) During this Mystical Experience, I drove to a tattoo shop and asked them to tattoo a bridge on my arm as a reminder. I saw I was to build a bridge in my mind whilst having this Mystical experience and it was the one clear thing that I pulled out of the experience. To me, in 2004, it was all about the bridge.

I had the most amazing spiritual awakening/mystical experience in my entire life. I reached what I would call enlightenment whilst I was on that bridge. Everything made crystal clear sense, intelligence of vibrational energy, the Universe, life. I felt so powerful, not controlling everything but I felt in partnership with the creator of this existence, this game of life. 'I shook the hand of God', is what I text messaged my friend whilst having this experience. It was the best experience of my life.

12) What is Enlightenment?

In Zen it's Satori.

In Hinduism it's Samadhi.

In Buddhism it's Nirvana.

In Christianity it's Visions of Christian Saints.

It is a Spiritual emergency for Stan Grof.

A Nature Mysticism for Ken Wilber and a Peak Experience for Abraham Maslow.

What is Enlightenment in the mainstream today?

According to science, psychiatry and society, enlightenment is insanity, bi polar manic crisis, acute psychosis and it doesn't exist and needs to be stopped with medications at all cost. What's happening to our world? This is very real and it's the direction we really need to be going.

Eveline "Maybe add in a section explaining how you discovered that you are using 70% of your brainpower. So what does that include and where did you read this or people will think you're just making up numbers."

David "I discovered it at a lecture given by a licensed psychiatrist lady called Robin, who had a spiritual awakening and she spent twenty years studying this experience to which she added that when a person is in that awakened state, we use approximately 70% of our brainpower. Her experiences were very similar to mine and I felt the same level of consciousness in that experience as I was never so awake in my life, reaching a level of consciousness that was never even imaginable before in my world."

Step 8

Astral Travel/Astral Projection

1) After I experienced many states of mind that were on the negative side of life and were taking a toll on my body, I found an answer and that was projecting my consciousness outside of my head and looking back at myself. During this process, the negative thoughts in my head stopped as there was no energy feeding those thoughts, my energy was outside of myself and I could just watch those thoughts die out.

2) I figured that I found the answer to depression and all those negative states of mind that were affecting so many people in my world. I later looked into this and found a subject called Astral Travel.

3) Astral Travel
I went to a course with my friend called Rachael. It was called Astral travel which involves letting your body and mind fall asleep but you, your real self stays awake and then you can leave your body and experience being awake and alive and existing whilst your body is asleep. It didn't work for me whilst I was doing the course.

But on the 9th of August 2004 at 3:30am, in my apartment, I achieved the Astral Travel. I went to sleep that night and just as I lay down, I said that I was going to let myself Astral Travel tonight. I woke up at 3:30am and came sliding out of my body, out of the bed and was expecting to fall on the floor, but I didn't, I just hovered above the floor. I was fully alert and I moved in an upward position and I saw myself lying in bed. I felt great and weightless but panicking at the same time as I was out of my body. I moved over to the bed and did a complete spin, how I was moving was by use of thought. I moved over to the window and I was thinking of going outside but then I thought what if I couldn't get back into my body and I have all of these stories to tell, my life would have all been a waste. I then moved back over to the bed and I lay on top of my body that was lying there and in a moment I was back in my body. I opened my eyes and jumped up and my heart was racing just realizing what was after happening. I grabbed a drink out of the fridge and I raced up and down the sitting room, trying to comprehend the experience and the possibilities it might have. It was truly an amazing experience.

4) There are a few names for this experience, Astral Travel, Astral Projection, leaving your body and same experience happens when you die. Astral travel is leaving your body and traveling around by use of your consciousness, you don't see yourself and your thoughts let you travel to where you want to be. Where as Astral Projection is leaving your body with the consciousness of your body, you can see yourself as your whole body whilst your physical body is also lying there on the bed.

5) When you do this, you don't want to look in the mirror or you'll get a shock. I have been told this many times and haven't done it. Your body appears distorted.

6) Where your mind goes, the energy flows. You can stay awake all day and live in this world and all night you can Astral Travel while your body rests and your spirit is free to roam, we can be conscious always or better still, we can be aware that we are conscious always.

7) How to Astral Project, you lie down in your bed and imagine popping out of your body and walking around your bedroom. You use your imagination to walk around the room and pay attention to detail, the feel of the carpet and the cold from the fridge for example.

8) Bring your awareness back to yourself and make the intention of yourself falling asleep.

9) Notice the affects, the ringing noises in your mind, the movements, the shifts, and the vibrations. Notice your body falling asleep and make the intention to get up and make toast or get a drink in the kitchen.

10) Once the ringing noise dies down, you get out of your body, you just get up as normal but as soon as you get up, your somewhere else.

11) We live in a world of energy, thoughts, vibrations and frequencies. There are many different dimensions all happening at the same time in the same place, all happening on different frequencies. It is just a matter of what level consciousness we are at to see that, or to experience that. When we leave our bodies, we are on a different frequency but still connected with our present reality.

When we die, we don't go 'up there' or 'down there'; we are everywhere and are able to move about freely. It's a lot like the Internet; you can't just go there, it's everywhere, just like life after death or Astral Projection.

12) After I left my body and had that surreal experience, I had many conversations with others who also experienced the same experience. Some of the stories where so wild, I never would have entertained only I had the experience myself and I never would have heard the stories, only I had this experience and attracted more like minded people. This experience I had is only the peak of the iceberg, it's the opening of the gateway to a whole new world. There is so much happening on that frequency, one has to see for themselves.

The Celts knew all about this and have written stories all about the underworld and many other stories of the symbolic worlds, that we in this present time still don't understand. But in the Celtic days, they used to leave their bodies and travel through the underworld, in what we call now Astral Projection.

Eveline "That's really interesting. I think there's another word for it as well. I think it may be lucid dreaming but I'll double check that now.

I don't think they are quite the same but lucid dreaming is when you are conscious of your dreaming state and have control over what you do and see."

David "Thanks! Next step?"

Eveline "Yes."

Step 9

The Illusion Of Time

1) Time was first calculated by the Earth moving around the Sun. The first clock was a pointed stone clock that showed the shadow from the sun. The shadow revolved around the stone plate as the earth rotated the sun. Time started off from the sun rising in each city. Time was always different in each city as the sun rose at a different time to the next city.

2) Time is different for everybody. The time changes depends on where you are in the world as it was first recorded by the sun rise and sun set. For example Ireland is nine hours behind Australia due to when the sun rises.

3) As the sun rises in the east and you start driving west, when the sun goes down in the west, you have gained more time as you have travelled more of the earth in a westerly direction having more day time than before. It depends whether you travel north or west will change the perspective of time for you.

Time also goes fast for people when they are busy and slow when they are idle. Einstein says that once we move through space, time goes slower than that if we stood still which was tested and proven to be accurate.

4) The most interesting part is, time is an illusion.

Everything that changes in our world is an illusion. The only real consistent in our world is our spirit, the soul, our true selves, our consciousness, our energy.

5) Life is Now – there never was a time when your life was not now, nor will there ever be. To be free of time is to be free of the psychological need of the past for your identity and the future for your fulfillment.

6) Time and Mind are inseparable.
You are the energy that feeds your mind and once that energy is in the mind, the mind believes the mind is real at all consequences. From the minds perspective, time is very real as time and mind are inseparable. When you step out of mind and become present – you get the best out of your mind when you decide to use it as apposed to being held captive in psychological time, draining your energy.

7) There is more of you apart from just your brain and mind. We are made up of our minds, our consciousness and our super consciousness. The perspective of our reality changes when our consciousness changes, the more we become aware of our own consciousness, the more truths we then know.

8) I was in Sydney Australia, finally had my life sorted out for a while and I needed to work out all my previous experiences. I was trying to work out how I could see future occurrences before they happened and how I could feel the energy of an event before it happened. I had so many questions about how is this all possible and I agreed to myself that it just was not possible, although I experienced it, unless I take something out of the equation. I took time out of the equation and all fell into place, every experience made perfect sense as it works of energy and how I could see future occurrences was because they were happening due to the direction my energy and vibration were heading. Once I grasped that concept that time is an illusion, a deep part of myself woke up. I crossed a bridge within from mind to consciousness;

I woke up and entered the very existence of the Powerful presence of the Now.

This was one of the happiest if not the happiest realizations of my life. Time is an illusion. All else faded away. I always exist in the Now. So many illusions became see-through. For example, we can always cope with the present moment but we can never cope with the future nor do we have to. The answer, the strength, the right action or the resource will be there when we need it, not before, not after.

The only time that exists is the present moment, in a timeless reality. This causes you to step out of the mind, which is either constantly thinking of the past or the future, it can't exist fully in the present but you can. You stop dishing out 60,000 thoughts per day and become fully aware in the present moment and there are levels to this awareness, which we'll get to. All things that truly matter in life – beauty, love, creativity, joy, inner peace, all arise from beyond the mind, from the true self, in the present moment, the Now, the only real reality that truly exists. All true artists create from their true selves, not their minds but from their consciousness.

9) We are born, we become children, we grow up, our physical body changes but we still have the same energy in us, our real selves, our consciousness, the watcher. This energy of ours only changes from experiences we have, we evolve through experiences that are all happening in the Now. Our level of consciousness changes as we move up through the levels, all happening in the Now, it's like one big game of hide and seek. We need to seek to find ourselves, to become more aware, to higher our consciousness through experiences, which is all happening in the Now. There is no other time only the present. We are spiritual beings having human experiences and everything we think we know using our five senses is an illusion, we are internal infinite beings in a timeless reality.

10) The present moment is followed by another moment which is the Now broken down into moments but one must get out of mind and into their consciousness to experience this. Anytime anything

happens, it happens now. All events in your life happened in the Now, the future is a possibility of the present and the past is borrowed from the Now, everything happens now. There is no other reality. It's always the Now.

In our timeless reality, we are energy moving through experiences brought forth to us by our thoughts in our minds.

11) Everything is happening Now. Energy, thoughts, vibrations, frequencies, in a timeless reality, all happening now is the really real world, past the illusions of our minds. We are energy that vibrates, we have thoughts that also vibrate and are made of energy and as we always exist in the Now, we move from FM frequency to AM frequency through the cycles of birth and death and can also tap into different frequencies by raising your consciousness in the present moment, not to mention Astral Travel where we travel consciously whilst our bodies are resting.

12) When you wake up to the Now, every cell in your body is so present that it feels vibrant with life and when you feel you're present in every moment, then it can be said that you are free of time. To be free of time is to be free of the psychological need of the past for your identity and the future for your fulfillment. It represents the most profound transformation of consciousness that you can imagine.

Eveline "That's fascinating, I never thought of it like that, but I still can't work it out, how it's possible that time is an illusion."

David "The more you keep thinking of it, the further away your getting as you're racing your thoughts in your mind trying to work it out. The only way to realize the illusion of time is to not think, to become present without thought, to cross a bridge within, from mind to consciousness and that's where the real reality lies away from the physiological insanity of the mind.

Interested in the Power of Now as next step?"

Eveline "Yeah."

David "That's your favorite step yes, from the start!"

Eveline "Sure is!"

Step 10

The Now And The Power Of Now

1) "We are very good at preparing to live, but not very good at living. We know how to sacrifice ten years for a diploma and we are willing to work very hard to get a job, a car, a house and so on.

But we have difficulty remembering that we are alive in the present moment, the only moment there is for us to be alive."

Thich Nhat Hanh

2) I once heard in poetry English class when the teacher was reading a poem, the poet had written, he was in the land of Nowhere!

Right there was a seed planted in my mind and the mind always ticked over trying to just grasp that but I could never figure it out. I also believe that my teacher just felt it was poetry and cannot be explained but that didn't cut it for me as it was clear the poet knew what he was talking about only it was so deep and in so few words, no one could totally understand unless they ended up there.

So for me, many years later over various reasons of experiencing different states of mind and a few attempts to end it all, I finally killed that racing mind and brought myself to a new awareness. I really totally connected with my consciousness, my energy and I walked around not in my mind but in an existence unknown to the majority of mankind in this day and age! This place is called Nowhere! And that really means Now-Here! The more you think of nowhere, the further away your getting, it's a state of being, not a state of mind.

3) I went to hell (state of mind) about four times in my younger years and came back out of it because of the love I had for family but I went to hell again and I had to give up on living, after four months of mental inflicted torture, my mind was wrecking my body causing weight loss and coughing up of blood and basically a torturing state of mind. So I gave up whilst listening to a U2 song by means of electrocution but I survived. Then I found something in between the thoughts of my racing mind and after another month or so after totally understanding the illusion of time, I ended up in Nowhere! Now, I tell you this, it didn't feel like it at the bad experiences but it was so well worth it at the great experiences. What's the point living a life in all sorts of states of mind and fighting with your mind to stay on top or remind yourself to constantly think positive thoughts to survive and never finding yourself, sorry but that just doesn't cut it for me.

The reality changes when your consciousness awakens, there is a lot more to this life than constant thinking and dulled reality and lack of energy. But now, after torturing experiences, I have found very easy ways for students to get to Nowhere, the only real existence there is in this world. It's totally mind blowing and gets deeper than that – soul awakening!

4) Releasing the energy within.
On the process of getting to the present moment, we must first release the negative energy within. Many of us have a ball of energy in us, which has been passed down from generation to generation also known as the pain body. This energy consists of negativity of some sort, whether it is angry at times, powerful negativity or feeling fear, this energy needs to be released instead of bottled down on every occasion that it rises up. When it's leaving you, it leaves out through

your minds eye, (your head) but it cannot leave if the mind is racing constantly. Once the thoughts in your mind are racing so fast, they act like a blockage that stops any energy within leaving us. It's almost like a lid on top of a cooking pot stopping the steam from coming out. Nothing is really solid, even within a rock, the molecules are moving so fast that it appears and feels like the rock is solid but the rock is made up of 99percent space. When your mind is racing with endless thoughts as this energy within is getting brought to attention, you need to stop the thoughts or you become unconscious and are controlled by this energy and end up in a fit of rage, fear, anger, etc. so you become present and find a state of being in order to release this energy and then when that's done, you are over that huge hurdle as you release the same energy that has been attracting all the bad experiences (same negative vibration) in your life. Then you can enter the Power of Now after surrendering and releasing that negativity within.

5) I lived in the Now for about six months, yes sounds like a contradictory, totally in Present – in between Thought, Timeless and yet I also say six months. We need some form of clock time to track lengths of experiences and no-time (total awareness in The Now) can coincide with Time. We can use time as a tool whilst we totally exist in big moments broken down to a lot of other moments. But time and mind that are inseparable is the mind thinking either in the future or the past and cannot exist presently.

I slipped out of The Now, as I wanted to get everyone into The Now but the bridge was missing, there was a big gap in between where everyone else was and the place where I became.

So I relived my life – tracking my past and writing about each experience, just as life presented to me, in hope to fill the missing links but somewhere along the line, I got trapped in my past, some bad experiences and I ended up writing about a whole load of experiences where I had a lot of personal issues surrounding them. This writing helped me to release and to overcome this baggage I dragged up and was then carrying. I did write about each experience and it sort of helped bridge the gap but it only led people to feel rejuvenated, it didn't bring them into the only real existence that exists. So, here I am, trying again!

6) In the Now, I look at the foundation. When being a parent, your soul job is to put magic, fun and great memories in your child's life. Fill your child full of great experiences, laughter and fantastic memories and when you finally get old, you'll have the love, the support and the happiness that every parent dreams of.

The Now is the only existence there is. Always be aware in the Now.

Look at it this way, have you ever experienced, felt, thought or done anything outside the present moment? Do you think that could ever happen? Is it possible for anything to happen or be outside the Now? The answer is obvious, is it not? Nothing ever happened in the past, it happened in the Now. Nothing will ever happen in the future, it will happen in the Now. The essence of what I am saying here cannot be understood by the mind. The moment you grasp it, there is a shift in consciousness from mind to a state of Being, from time to presence. Suddenly everything feels alive, radiates energy and emanates Being.

7) The Now is a place in-between thoughts. Buddha made it and many more people made it too. Religion has a way of blowing things out of proportion and religion restricts people by making them believe this place, this level of existence is only allowed for the Buddha for example. A belief keeps people restricted. Religion renders people powerless, Spirituality gives people power and that's the difference. I'm just trying to give people back to themselves and have no interest in organized religion. I'm beyond that and always was.

8) The Stop – Start mind and Infinity.
Once you cut energy in half again and again, as there is no smallest, there is no biggest! The mind can't grasp infinity! Once you grasp infinity, your consciousness is waking up. The mind is a tool for you to use. Your energy is full of intelligence and it feeds your mind. Use

that tool and manifest, I'll show in the next few Steps. I have been 'imprisoned' in my own mind in the past and that leads to total and utter self-destruction! But now I have learnt tactics that fixes all states of mind, only from previous experience and the next section will show you how.

9) Life in the Now, feels so great it is hard to put into words how great and real it is. For starters, all your senses become heightened. You feel energy, powerful healing energy coming to you in waves. Your hearing becomes so perfect, you could hear a pin drop, your sight is perfect and you can see the perfection of each object, fully enhanced in colour and so alive. You can see each drop of water out of the shower, your vibrations are higher than that of normal and you're so aware of what's in your environment, your surroundings. The feeling of the water on your body whilst having a shower is amazing, so comfortable and cleansing. The taste of water is so incredible and feels so good to the body. The taste of food is so rich and enhanced, you feel so alive like never before. We have five senses that we use, hearing, sight, touch, taste and smell but in the Now, alive in the present moment, there are also levels to this which I'll explain after but we also have our sixth sense wake up as our minds are not in past and future thinking, we are awakened and our intuition is there, fully active. We can feel all around us and other people's vibes whether positive or negative. When a person comes to you, you can totally sense whether that person is coming from a position of negativity or of some form of love, positive energy. You become so aware of everything, everyone's energy and every vibration around you. When people are talking, you're so aware and hear every word and the space between every word and you're ready to answer before they have even asked the question as we don't have our energy dished out to 60,000 thoughts, we are fully aware and present in the Now and awake to the same essence that is in us all. There is a vast intelligence beyond thought. Also, we can feel future events coming to us as we can feel the energy and vibration of that event.

This reality is how life should be only our minds that live in future and past thinking constantly gets in the way and dulls our reality and our senses as our energy is dished out into 60,000 thoughts per day.

How to get into the Now and the Power of it. The other way of getting away from Past – Future – High Expiations and Imprisoned in your own mind is to Stop what you are thinking and use your senses.

List 3 things that you can See!

List 3 things that you can Hear!

List 3 things that you can Touch!

List 3 things that you can Smell!

And List 3 things that you would like to be Tasting!

This brings you into the Now – it keeps you present!

Practice makes perfection and consciousness gets awakened! Once in that awareness, you're in a state of being and not a state of mind as you become aware of your true self. Once in that awareness, naturally your sixth sense comes into play and then you have found yourself, you own your life and your world and it's all fun times and exciting times ahead as your sixth sense lets you prepare for every big event that is going to happen. You become connected and can feel it all. This is just the start to awakening and it's infinite! The question is, how far down the rabbit hole do you want to go?

10) One of the consistent things during my life of moving from different states of mind to different levels of energy, was that, I always attracted the like energy people out there, people who were on the same vibe as I. Whether is was depression, or anger, or the mystical experience, or having the release of energy from within, or the awakening of the Now. It is only when my energy is vibrating at that speed, do I meet people that I never knew existed, people who are also on that same vibration, who I never would have met before.

When I broke free of the negative energy within and became fully aware in the Now, totally understanding the illusion of time, to my amazement, I met people on the same vibration and we were all one group of timeless people.

We were in the Now, not held down by the minds endless thoughts and experiencing pure awareness. I've attracted many people in this vibration that I had no idea ever existed. It became totally mind blowing but I'm totally in the Now so 'mind blowing' gets moved to 'soul awakening'.

We all had a chat and when I explained experiences and illusions from my end, one of the other gents said that there is a good book written called the Maya which means illusion and that it's held in a museum somewhere, locked away and no one can read it. The other gent reckoned that it is his mission in life, to steal that book and they all agreed that this reality is known, many before have experienced it but that the governments and religion control people. I was just amazed that this energy attraction, although I saw it happen throughout my life, still it was happening when I attempted to end my life and survived and woke up to a new reality, that there were others there too, I wonder what they went through to get to the Powerful presence of the Now.

Now there is a spiritual rule out there and that's let each man, woman and child walk their own path. But I feel that rule needs to be changed rapidly the way the world is going these days.

11) Liam, my other student, who I showed how to get into the Now and to live the power of it, successfully passed the test and achieved it. He was having epiphany after epiphany and he eventually vomited because he received too much information at once but he achieved the Power of Now and said that he became three levels deep inside the Power of Now (experiencing levels deep of awareness in the present moment, seeing three reasons why each moment occurs).

But later he asked, "David, can you help me out at some point with a problem I've run into, I would love to talk to you about it but I'm out tonight and leave for New Zealand on Monday so I might not get the chance.

I made a decision, Thursday morning, to begin shifting my life back into the present, as you made me realize just how absent I was being. I successfully did this and had two hours sleep and worked two shifts, still charging at the end of the second day, no problem. But then I crashed horrifically and had to fall into a slumber for seven hours last night and another three hours during the day today and I feel as though the more I sleep the harder it is to get back to where I was (living a little more presently) but the further away from the present I get the more I need sleep. How do u focus? And stay in the present? How does one maintain this for a long period?

Even a little help would go a long way because I intend to pillage New Zealand with my newfound knowledge and rupture the earth under the skiers and snowboarders that think they can challenge me on the slopes.

Whenever you get the chance to reply, cheers.
Liam."

"Hi Liam, Okay, The NOW is the power!

Write NOW on paper and hold it to a mirror and what do you see? The level you're at in the Game of Life!

Answer to get into the NOW.

Use your senses – not your mind!

Need to let go of the Ego mind and break free of it.

The Now.

Here's the trick.

Using your Senses.

List 3 things that you can see,

3 things that you can hear

3 things that you can touch

3 things that you can smell

3 things that you can imagine tasting.

Also, Look at time as sideways, horizontal. Then you become the watcher. If vertical, you relive your past and bring into your future.

Yo mate David!"

"Hi David,

What if I attempt to stop thinking, try to absorb myself in the present through the senses I've got but find myself so unable to for more than ten minutes because my vision is fuzzy, my hearing is dulled, taste buds are stagnant, nose is blocked and my fingers are numb?

This is an exaggeration but today I had my eyes open but wasn't actually focusing on anything I was seeing, no matter how I forced them open. I am about to go find a paper and pen and will also get back to you. Now I've had a nap I'm finding it easier to open my eyes but that's just one step of course but I don't want to have to nap to be able to do this!

I love what you have been saying about not thinking and instead playing the game, it's quickly becoming (I would say obsession but that implies overthinking, which is the opposite of what I experience) a way of life. Great things happen living in the present, it only took me 48 hours to find this out but no matter how hard I try I feel LIMITED by my being rough. About to go get a page.

Now is backward! Not sure if I understand but I'll tell you what I see just in case you wanted to know whether I do. My past self wrote now in the now to the future self, looking in the mirror and it became backward. Am I on the right track? Is this significant?"

"Hi Liam,

It says you WON! Life is a game, Get in NOW, You WON, Go 7 levels deep in NOW and you'll find me there.

Waiting for you!"

"Hi David

I pulled over in the middle of driving to reply.

Wanted to let you know what you taught me so you might be able to grasp just how much of a teacher you are to me!

List of realizations.

Written on a page it reads 'won', but the letters are reversed by the mirror. The real win is when I am in the now, I suppose.

Time isn't as we perceive it, just because I made backward progress one day, doesn't mean my potential has gone backward, I'm just experiencing a new day, a new segment of time, not backwards or forwards, just a different point in the loops and loops that are time.

I'm thinking about this a lot and I want to stop but one must think to learn and one must learn to apply those skills. Now I'll stop, because I have some skills to apply. Love ya mate."

12) The Now is The Answer. My most resent adventure into the Power of Now, was just that, very powerful. I used the list of 3 things to use my five senses and to get myself present. The first attempt did not cut it but practice on that makes perfection.

After doing it many times, I became very present. Once I was present, my sixth sense woke up and I intuitively knew what was about to happen before it happened. It worked perfect with big events as I could feel the energy of the event, I could feel the collective energy that would surround the event. During this period, not only did I become totally present and naturally using my six senses to navigate in this world but also I became deeper in the Now. I realized that there are levels within the present moment and a power to be harnessed. As I practiced becoming more and more present and listening to music that stimulated my mind, I moved more into the Power of Now. I was getting what felt like waves of energy coming into me. It was powerful and also healing as this energy that was coming into me, waves of it was healing my body and I became healthier and happier. My body was getting itchy all over like it feels when an injury is healing, my whole body was feeling like this, my whole body was healing, energy was going to all cells and it was feeling so great. Once in this Power of Now, as you move more and more into it, you get more and more energy and you don't need as much food as your getting the energy from within, your connected and grounded. I needed less sleep and basically I slept when I was tired which occurred every second day. I moved levels deep within the Now and I could see reasons for each moment that occurred. Everything became so alive and every moment had reasons as to why it was happening. The deepest I went, seven levels deep inside reality, the game of life that most people don't even know they are playing. Life is a game of awakening and of enlightenment. I became so full of energy and my body healed from any physical injuries that I had and oxygen went to every cell in my body and alive I felt in each and every cell of my body. Each moment that took place, to me consisted of seven reasons as to why that moment was occurring. I could see all the people, identified with their minds; they were their minds and were far from this present reality. I became so in-trenched in the Power of Now, that I became many levels deep within the present moment, accessing energy that I never knew was there and witnessing each moment for the many reasons as to why it was happening, I was so alive and so aware and so very awakened.

The inner essence of nature, people and beauty shines through but it only reveals itself to you when you are totally present as your presence and it's presence are on the same level of awareness.

People's energy and vibrations were raised when I was around which made many people so happy, they felt great and so alive and so intense whilst talking to me. They became very much aware and awakened and exhilarated. Some of them sang songs at karaoke and one gent was saying, "I just can't stop singing, I've never felt so happy before in my life."

My energy was vibrating so fast and I was having the best natural experience of my entire life. I was writing three separate books, running a carpentry business and looking after my family and still out there with friends and meeting many new people, basically I was harnessing powerful energy that I never harnessed before. There is a vast intelligence beyond thought and you receive constant inspiration from that energy source as you become your own energy, you become connected to that inspiration and basically you become it.

I could feel the world's vibrations, the weather changes that were happening everywhere, I could feel the earth's energy and vibration. I became totally connected, the ego collapsed long before and totally present and awakened I became as I became my consciousness. There is no limit to this awareness; this power goes on forever, it's just a matter of how aware you can become and how deep within this reality, the Power of Now you can go.

The more you become aware of your own consciousness, the greater the experience, the greater reality is, as the truth becomes an inbuilt wisdom when you access the vast intelligence that is there waiting to be discovered.

I, life and the world needs more people that can harness the Power of Now, for it's the answer, it's on the path to enlightenment, it fixes all problems, whether states of mind that are wiping us all out or making us sick or causing health problems, as we receive powerful energy and healing energy and over all it is the most powerful experience ever.

Eveline "That explains a lot, I always wondered how people were searching to be enlightened and the best way to go about it, as meditation doesn't work for me."

David "Yes, meditation didn't cut it for me either, I hope you get there – it's what life is supposed to be about.
Next step?"

Eveline "Next Step."

Step 11

Knightsbridge 1919

1) I lived in a place in Sydney and I was working two jobs and no matter how hard I worked, I still was struggling with money and bills and rent. So, I changed my thinking and I figured it was about time that I enjoyed life and lived in a place with cheap rent and had some good friends and lived life for a while. Once I changed the thoughts in my mind to a new level of consciousness, as I realized that what I was doing was not working, I then attracted new opportunities into my life. It started off with one person saying to me in work that I have the same accent as another person he plays football with. This bloke turned out to be a friend of mine that I had not seen in fifteen years. Soon, I met him again and I moved to a new town and new accommodation where he was living at and that was free, a new job that sorted my visa and many great friends around. My whole world changed just by changing my frame of mind and a whole new reality was brought into my life over the next week or two as I knew what I needed in life and I focused on what I needed and felt I deserved it.

After enjoying myself immensely with new and old friends for many months, I then decided to get back into writing and working out what I was experiencing after I tried to end my life, what was that reality I got a good glimpse of, that was in between thoughts and how could I see the future and then it became real in the present. There were many questions I had in my mind and I couldn't stop trying to work out how my experiences were possible. Then when I was explaining to a friend that the only way my experiences could make sense is if I take something out of the equation, I removed time from the equation and everything became very possible with energy and vibrations and how to see the future. Once it occurred to me that Time is an Illusion, I crossed a bridge within. I already crossed this bridge with a few attempts of ending my life but even that was just tasters of the experience of Being until I realized that time is an illusion. Then, my world changed as I moved from living in my mind, to living in my energy that feeds my mind. I entered to the Power of Now.

I moved from where I was living to a new apartment called Knightsbridge 1919, that was written across the front of the apartment and many experiences happened there.

2) I became very much manifested in the present moment. Everything felt so good, all senses heightened and everything became so alive. Energy and Vibrations and Frequencies and Thoughts are all that really exist. I could feel energy around me, I could feel who it was, two energies on AM frequency where in my apartment and one was called Florence who was my Grandmother and the other was called Vincent who was my Granduncle.

There was a girl I was hanging around with called Rachael who was more or less on the same vibe as I was on and although I could feel energies around me, she on the other hand could see and hear what those energies were doing and saying from the other frequency they were on. Rachael had this really sensitive hearing and always complained about any party's that happened around the area, due to the noise but she had a gift and she could hear other spirits talk to her. So between the two of us, we could see spirit, hear what these people who passed on were saying and feel exactly who they were and where they were in the apartment, it was all very interesting. I thoroughly enjoyed this experience. Every energy that I felt on AM frequency, felt really good, uplifting and I could feel and sense who that person use to be, an inbuilt knowingness that comes to you when you open your sixth sense, fully present in the Now.

3) As I became totally in the present moment, the endless thoughts and chattering stopped within my mind. The energy within me, the negative body of energy that I used to bottle down was coming up in me and I was very happy to let this happen. But whilst this was happening, it was also bringing Rachael's negative energy to the surface and she didn't know what was happening to her but I explained that this is so good, this pain body energy is finally leaving me. Rachael was reacting with this negative energy and all sorts of childhood family problems were coming to the surface in her. We needed to get away from each other to let this happen naturally as racing thoughts in the mind stops this energy from leaving. I moved out from my apartment and went to a hotel to enjoy this process. I said out loud, "I surrender" and I felt an immense weight off my shoulders and I let go. I allowed this negativity to leave me. It was almost like it was a solid weight leaving my body and going out through my head. It took about two days for this whole process to happen and I felt so happy, there was hope in my life and I have been released from that negativity that was fuelling my minds thoughts for as long as I could remember. I was on an emotional roller coaster for so many years with highs and lows and making it to this experience was truly incredible. I became free and lived beyond the negativity of the world as there was no more negative energy attracted to me, as there was no more negative energy in me.

4) I went to the post office across the road from where I was back living at Knightsbridge 1919 and there were four people serving. It didn't bother me anymore to be in this place, the bright lights were not drawing the negative energy in me to attention, as the energy was released, everything was just so clear in the Now. The people serving were a man who was dressed like a librarian with white hair and glasses. There was a woman who was dressed in black with a veil over her face. There were two others also dressed in a religious way, a Jew with a hat and a Muslim. It was so clear to me, all history is past and held in the mind. Their religions were belief systems passed on through generations and this was what they believed themselves to be. Basically, all in the Now, they we're four people dressed up in different ways and none of them saw that they were in the eternal moment, all the same stuff, all from the same consciousness. They were all completely separated from each other and the separation was their history, religion, place

of background, country of birth, it all became so clear. Their beliefs in their minds had them all believing that they were so different but there we all were, in the post office, in the one moment, all made of the same stuff that makes us all One. Everything became so clear to me as I was awake and knew the same essence that gave their minds life, was the same essence that I had woken up to. Everyone was in his or her minds; no one was awake in the moment, only me. It was like a play, actors all in the one room but it's called real life and the actors actually believe all that they had been taught as their beliefs in their minds were so real. The mind is like a computer with a series of programs running, constantly thinking of many things at once. But they were sleep walking, didn't see the truth. The truth was, I am my spirit/consciousness now, I could see that their spirit/consciousness and my spirit/consciousness were from the same spirit/consciousness; we are all from the same essence manifested in this game called life.

5) You know when you have an 'Aha' moment and what was happening in your life makes perfect sense all of a sudden? Well, when you totally enter the Now and go levels deep within the Now, all around you and all of your life becomes an 'Aha' moment, you totally understand and see so clear as to why everything happens. You become all knowing in this life and even that is just a blimp on the radar, you know of a time before and you feel that finally you have come home to yourself. You previously thought that this could only happen when you die but you are experiencing this whilst very much alive and healthy and well. It is a truly amazing awareness and experience.

When you get into the Now, you begin to realize that there is a massive infinite empire of intelligence beyond thought and that thought is only the tip of the iceberg of that intelligence.

6) Rachael and I were talking about religion and religions don't ever say what it is, only that it is the end of pain and suffering like in the Buddhist religion for example.

The Now is a place in-between Thoughts. Buddha made it and many more people made it too. Religion has a way of blowing things out of proportion and religion restricts people by making them believe this place, this level of existence is only allowed for the Buddha for example. A belief keeps people restricted. Religion renders people powerless and Spirituality gives people power, that's the difference.

There is a missing link though. It's Nirvana, it's hard to put into words, it's every sense awakened and you become awake within thus leaving you connected, in partnership with the intelligence vibrational energy that created this game called life, so you become all knowing. You may not know how to play an instrument or to do mathematical equations but you become all knowing on a conscious level, you experience truth that is beyond the minds comprehension.

What most people don't realize is that together or individually we are what men have called God, for use of an analogy, we are the drops in the ocean and the ocean is God. Each drop holds in it the same particles as the ocean. It's only we are trapped in our minds, that's the illusion!

For use of an analogy, we are the mobile phone without Wi-Fi when living in our minds, once we cross the bridge and become manifested in the Now, awakened, we get Wi-Fi and are connected to the all knowingness that is out there, which is us, our deeper selves.

7) I became totally connected with the Earth's energy and vibration. I could feel the weather changes.

This sixth sense in me was totally awake and I knew from within what was going to happen before it was to happen. There were so many changes in the weather this period of 2004 and I could feel it all. On one particular occasion, I left the apartment and I looked up at the sky, the sun was shining but I could feel a change in the day. I went back to my apartment, Rachael was at work and I could feel it coming, I could feel that a big snowstorm was about to happen which was totally ridiculous as it does not snow in Sydney. The previous snowfall in Sydney was about 100 years prior. I had a CD in the apartment with Christmas songs on it and I put it in the CD player, playing Christmas music. I took out the camera and took pictures of the blue skies so I would have before and after pictures.

I then texted Rachael and said, "Happy Christmas." She messaged back and said, "What are you talking about, it's only September?"

I replied, "Wait for the Snow!"

Then it happened, snow fell in Sydney, Australia. Not just little snowflakes but big heavy balls of snow came down from the sky. Rachael rang to ask what was going on and I said, 'It's snowing, it's just started, can you not see it yet?"

Rachael said, "It's not snowing, it doesn't snow here . . ." Then she yelled, "Oh my God! Look at this, it's coming over, people are running under bus stops, the snow is huge! Oh my God!"

I had the camera and took more pictures; snowballs were coming in through the window. It was amazing. The minds of the world with all their analyzing on weather maps and forecasts could not see this coming but by being connected within the Now allows you to feel all of this before it comes to play.

8) Astral Travel
I went to a course with Rachael. It was called Astral travel which involves letting your mind and body fall asleep but you, your real self stays awake and then you can leave your body and experience being awake and alive and existing whilst your body and mind is asleep.

On the 9th of August 2004 at 3:30am, at this apartment called NightsBridge1919, I achieved the Astral Travel. I went to sleep that night and just as I lay down, I said that I was going to let myself Astral Travel tonight. I woke up at 3:30am and came sliding out of my body, out of the bed and was expecting to fall on the floor, but I didn't, I just hovered above the floor. I was fully alert and I moved in an upward position and I saw myself lying in bed. I felt great and weightless but panicking at the same time as I was out of my body. I moved over to the bed and did a complete spin, how I was moving was by the use of thought. I moved over to the window and I was thinking of going outside, but then I thought what if I couldn't get back into my body and I have all of these stories to tell, it would have all been a waste.

I then moved back over to the bed and I lay on top of my body that was lying there and in a moment I was back in my body. I opened my eyes and jumped up and my heart was racing just realizing what was after happening. I grabbed a drink out of the fridge and I raced up and down the sitting room, trying to comprehend the experience and the possibilities it might have. It was truly an amazing experience.

9) Once I became totally manifested in the Now, so present, I asked life, the Universe, my guides to give me paperwork on this reality that I'm in, as I've so much to write already. I was writing a book called Life is a Game and I felt that I had too much to write about, by putting in the missing links as to how to get to this awareness. I didn't want to just write about how deep one can get into the Now and how it's on the path to enlightenment if the links on how to get there are missing. Later on as I was explaining to a friend in the building I was living at, the level of awareness that I'm at, he handed me a book and said did you ever read this book called the Power of Now by Eckhart Tolle?

I later looked over the book and said to myself, "Perfect, this is so much about the Now and the Power of it, this reality is written." Highly recommended book to read!

10) I was in work, working away making some money and I figured that I need to get everyone to this level of consciousness, particularly family and friends and the only way I can do that is by writing a book. I had a clear thought in my head saying that I should not be working, I should be writing this book. It went like this "Universe, listen to me; I need to write this book and not work anymore."

With that, I had an accident and broke my chest bone. I was not too bad that day but in the morning I couldn't breath and had to go to hospital. Rachael came with me and she was making requests to the Universe saying that she needs money and she put so much energy into these requests to the Universe basically demanding money. Anyways, we got to the hospital and when I was getting looked after, Rachael was in the waiting room. She said she was so present in the Now and then some security men brought in this woman who was very angry and emotionally unstable. Rachael said it was like something out of a movie as the woman broke free from security and ran to Rachael and attacked her violently. I could hear the screams from the woman where I was and ran out.

The security took the woman away and Rachael was very shook up.

Just as we were leaving, a nurse came after us and handed Rachael a victim of crime claim form. The money was to be on the way that Rachael was requesting for.

After learning the hard way for the two of us asking the Universe to be off work or to get money, we then came up with, when asking the Universe for anything always say at the end, "Under the Grace of God in a perfect way." Otherwise we get what we ask for but many times not in a pleasant way.

11) Knightsbridge 1919

My Grandmother was born in the year 1919 and the Bridge was what I was trying to write, to bridge the gap between the material and the mystical.

Rachael could see spirits in the apartment and was getting very edgy over it. I could feel two separate energies in the apartment and I knew by the feeling that one was my Granduncle Vincent and the other was my Grandmother Florence but they couldn't communicate with me.

Rachael didn't know who it was but said, "Did you see that, someone just walked past in the kitchen?" I said, "I know who it is!" Rachael was standing at the hatch area between the sitting room and the kitchen where there was a small glass bowl on a tripod that was used for candles. It jumped out of its position towards Rachael and landed on the floor. Rachael with her new high-pitched voice was freaking out and saying, "Oh my God, did you just see that, what's happening here, who is this?"

I was laughing, as I knew it was Vincent but also amazed as to how he could move things. I thought this was great as I was asking Vincent to talk to me. I said before, "I have so much to write Vincent but nothing on you, I have a whole empty folder here for you, talk to me will you? Your a Champion of a man."

Vincent was playing with Rachael, first the tripod, then her keys were locked in a drawer, seriously, the drawer just could not be opened anymore. Rachael was uneasy over all this but she told me that she could hear spirit talk to her and that she has done it before. Rachael was from New Zealand and she was very in tune with the Mystical side of life. I explained to Rachael who was in the apartment and soon enough Rachael sat down and said to Vincent, "You are David's family, tell me about David."

12) Vincent said, "David wants to be a teacher, will be a teacher, is more sensitive and passionate than me. His sphere much broader, his mind is too. He could teach me about the world if we sat down for a chat. I love to listen to him when he is inspired. He is a visionary and has a dream. He must continue and will create a beautiful set of new teachings. He is my big brother, I look up to him. This cannot be explained but it is so. We are connected from a time before when we were friends but I always looked up to him for guidance and leadership. He is a true leader and inspiration to the world.

He is my truth seeker, a champion in the world, adventurous and brave, more alive than I have ever been. He will create gifts, which no one has seen before but eyes/minds will see differently after. Mr David, I love you for what you do, is all divine and full of love, you are my champion too. The veil is lifted. Don't stop, believe in yourself and let yourself shine. It will be magnificent and so you are too." 9/10/2004.

Eveline "That's fascinating, I had an experience where someone died about two years ago, as he crashed a car into my fence at my house and flew out the windscreen and died instantly. I was the first person to find him as I ran out and called 000, I also felt that he had no pulse and was a total wreck. There's a small alley way next to my fence that connects my road to the main road and I have to walk through it sometimes and every time I do, especially at night, I feel someone else is present there with me and I have a really strong feeling that it's the man who died a few years ago."

David "So this feeling presence around is a common occurrence for many of us, that's great. So many people are aware of this and it's all on the path to awakening."

Step 12

Manifestation

1) Nothing has, past present and future ever existed without first a thought.

2) The quality of your consciousness at this moment is what shapes the future – which of course, can only be experienced as the Now.

3) A Manifester is someone who can attract and place into his or her lives what they focus on.

4) Once you get yourself connected, there is a symbolic world out there. It comes to our awareness once a shift in our consciousness takes place. Also a question you have on your mind will be answered by either the lyrics of a song playing or by overhearing a conversation or by conversations on the radio, the answers are drawn to you when you are in synchronization with life, living in the Now.

Manifesting starts first with a Thought. Then you build up upon that thought by thinking of it every so often and reminding yourself of it. Then you envision yourself in the situation manifested by that thought. You hold onto the vision and as you are sending out thought which is energy that vibrates, let the forces go to work and the vibrations and energy get attracted to that thought, that vision and soon comes into your reality. This happens a lot more than people realize, it's mainly memory loss that gets in the way.

5) When I was in secondary school, on my last year which felt like prison to me, I was not happy in school or at home in the final years of school. I constantly thought about how unhappy I was and all of the things that were making me feel unhappy until one day, it came to me, to send out thoughts on what I do want in my life, not more and more thoughts of what I don't want as where the mind goes, the energy flows. So I changed the thoughts in my mind to, I would love to live in a timber house where I would feel good, away from my town as I needed a break and with a mystical girl who is into the comforts of life and looks great. I thought up of a place that would make me feel comfortable, as school was making me feel very uneasy and a timber house, warm, with incense burning and overall a very enjoyable experience came to mind.

Once I changed my thinking, created new thoughts on what I needed in my life, down to detail and then visualized it, that's the important part, to visualize it. Then energy began to shape.

I went out with friends and ended up in a nightclub. I met this girl called Sandra and to me, it felt like a mystical experience when I saw her, all else was a blur but she was totally in my sight and we both connected. We chatted and danced. It was a great night. Then I was back in school, sad again but this girl got my number off one of my friends and rang me. She was older than me and when my friend told me she was going to ring me, I had to tell her I was at work, not still very young and at school.

We later meet up and within a week or two, I moved up to her house for two weeks over the Christmas holidays. When I was up in her house, I had a moment and realizations came in, her house was away from the town, it was made of timber, lovely and warm and even had incense sticks burning. I was manifesting my reality yet again.

Everything happened so quickly, but I was amazed at how all these circumstances took place, energy shifted, all to match what I was visualizing in my mind. A manifesting experience.

6) Manifesting occurs when you focus, you keep imagining the same outcome over and over in your mind, you put energy into it, where your mind goes, the energy flows, you repeat that thinking until you can envision it, once you have that picture in your mind and you can see it, energy around you is starting to change shape, it's starting to adjust to your frame of mind, to your vision.

Once you have the goal, the vision in your mind, energy starts shifting and new events take place. All comes out in the wash just fine. The vision soon becomes reality!

Manifesting works best if you get into a zone by either driving on a long highway or by finding some sort of place of solitude and listing to a song over and over again, getting into a zone and then sending out thoughts that come from your true self, though your mind, with no distractions and let them loose. Manifesting occurs when you visualize this happening, when you visualize the outcome that you are looking for in your life.

Don't keep thinking how it's going to happen, just visualize the goal, energy then gets shaped and molded and brought into your reality.

I was single, living in Australia on my own and went for a drive, listened to the same song that allowed me to go deep within myself. Spent six hours driving round a small town, deep in thought bringing into my reality a beautiful girl, that I am attracted too and all the best personalities about the girl. I pictured my future with this girl as my life long partner. Six hours spent manifesting my perfect love of my life and time well spent!

Within the same month, (I could still feel this girl coming to me as I held onto that manifesto on for so long) a friend of mine said come around for drinks. I did that and some time later his wife said, "My sister is looking to come to Australia and is available." She showed me a group photo with her sister and with many other girls in the picture and I said, "Give me that girls number if she's available." (The amazing attractive looking girl in the picture) She got the girls number called Vivien whom I thought was way too classy for me but we text messaged each other and soon enough she came to me from 400ks away and the rest is history. She was everything I manifested and so much more as our minds can only think up so much, a lot like that mystical experience, Vivien and the mystical experience became so much more than the mind could possibly comprehend. We live happily with our daughter on the Gold Coast of Australia in a nice big house!

7) It's all about being rather than doing. Most of us run around doing things all the time but it's by being, is the point of attraction. Once you move into being, you then use your mind as the tool that it is and send out a thought on what you would like in your life in the pre-sent moment.

8) You need to listen to your intuition. It's the voice inside that's not from your mind but from your true self, your consciousness. And the only way to hear your intuition is by being present in the moment. That is the key to opening up your intuition and it knows all.

9) I have been playing with thoughts and energy since I was a child. It worked each time and I knew it was a level of consciousness, a level of awareness that I was on but I never told anyone how it works. It works by thoughts and visualizations. But the question arises; do we have choices in life? The answer is yes and no.
Can't really have a yes without a no, just like can't have a hot without a cold or an up without a down for one makes the other exist. There are always choices but then again there are events that have to happen, some personal and others world wide at a certain moment. We have to trust in the bigger picture, which can be understood by numbers. This game of life has existed, exists and should exist in the future. Within this game of life are seven levels of consciousness and this bridge being the higher level along with the Power of Now.

10) How to make, shape, attract and manipulate energy!
We have 60,000 thoughts in our heads per day. That is the same as having our energy dispersed into 60,000 different places per day.

Turn those 60,000 thoughts into One Powerful Thought and you can manifest before your very eyes. Or, think the same outcome over and over again, with a clear focused mind and you start to shape energy.

You can also start with a dream over constantly thinking about your future, then you move onto a belief. Now beliefs are very powerful, after that you move to a place called Knowing. Once you know something, you attract that all around you and within a day you could have 60,000 thoughts justified about this Knowing. Play the Game – it's here to be Played! That's a good way of turning 60,000 thoughts to One Thought as all thoughts are on the same vibration of energy. This means manifestation!

11) Imagine when you have a child and the baby is born, your minds thoughts from past and future stops for a moment and you enter an amazing moment! Or for most blokes, scoring a goal on national TV, or seeing a naked beautiful lady, or for a child, opening a fridge door and the fridge is full of chocolates, or 9/11 when the world stood still, consciousness awoke, well the mind stops and an awareness is brought forth for a moment. That's you connecting with your true self for a split moment. All there is in life are big moments – broken down to smaller moments but the mind confuses all that. Ok, let's say you see a sunrise or a beautiful sunset, you then enter a moment, well once you stop Time and escape from Mind, you move into your energy – your real self and everything changes, as you don't have a rare moment every once in a while, your total reality is moments! You see that you are really you – you know that you have come home to yourself, you connect with your consciousness and see that your mind is a tool to manifest the reality you would like in your life, you see it's all a game but you have been living in a dream inside a dream for so long and that you are so happy you have found your true self. The more you know about yourself – the less there is to know – full stop!

When in the Now, fully present, connected to your true self – this is also infinite as there are no limits to the levels deep you can go within the Power of Now. But those moments of inspiration that you used to have once a year or a month if you're lucky, that becomes your day-to-day reality. You are totally connected to what you once classed as a deep moment, you live in a moment that has other moments and in the over all day, it's a bigger moment. You see time is an illusion and all that exists is the present. You have constant feelings of euphoria, as your minds thoughts are not affecting your emotions, you're deeper and more aware than that and use your mind as the tool that it is to manifest very easily. Then my friend, not even winning the lottery would make you as happy as this as it is totally the answer to everything, I mean everything!

12) Now manifestation can occur in a timeless reality on the spot. 60,000 thoughts in your head into one powerful thought! This is how it's done. You become conscious when you cross the bridge, awakened from a dream state and free of the physiological need for time. You use your energy, which is connected to the vast realm of consciousness. You get intelligence from that place and you hold onto that which you would like to bring into this physical world, using your mind as a tool, you create pure thought using your whole mind without any other thoughts in your head as you are your consciousness, you don't live in your head anymore, the mind becomes the tool and you send it out. It feels as though your whole body is creating this thought and in a sense it is, it's your whole consciousness that is manifesting. This is how pure manifestation comes to play. Thank you for crossing my Bridge.

Eveline "Thank you so much for inviting me to cross your bridge! I found this book really interesting. It allowed me to see the world from a more spiritual and intellectual perspective and gives me insight into things that we don't process about the world. It shows the world and us from behind the scenes and shows us just how much is unknown to us about our thoughts, energy, vibrations and much more. I'm quite excited to see how others will find this book as this bridged my knowledge to understanding things far beyond my own thoughts. Different and brilliant!"

Closing Paragraphs

I am Life Path Number 7 – I am Energy – I Vibrate – I have Thoughts – I live on FM existence called Frequencies – I love Technology, Science and the Universe and live for it – I had a Mystical Experience – I can Astral travel – I live in a Timeless reality – called The Now – I used to live at Knightsbridge 1919 with my Grandmother and Granduncle and for a hobby – I Manifest!

Each subject is infinite!

The choice is yours – how far down that rabbit hole you want to go?

If there is a higher power – It's an Intelligent Vibrational Energy that uses a tool called Frequencies. It is Consciousness.

Why is enlightenment so good?

Well, our every day-to-day experience leaves us with thoughts that can leave us shaken up or stressed or worried as the thoughts affect our emotions, as our emotions are a reflection of our thoughts. When you start waking up, you begin to understand that we are energy, we live on forever, that we can't be damaged or harmed. Our consciousness gets discovered to these new truths that come from within and soon enough we have no fear or any negative thinking and also experiencing that it's always Now. There are many more illusions to be dropped and many more truths to waken up too but once we do, our thought process changes and we become fearless with positive thinking and these thoughts with total present awareness are what has us enlightened and feeling fantastic harnessing a power that once never existed to us before. At the end of the day, it's the level of consciousness that we're on that allows us to win the game of life.

Consciousness.

If you ever wondered how is it possible how people connect with life after death, how people see the future, how fortune tellers can be spot on. Once you realise that we're all part of the one consciousness, you then see how this is all possible. Nothing is separate, we're all connected.

The Bridge to Consciousness.

Once you're over this bridge, you should be able to fix your crown in the mirror and soldier on in a new world of infinite possibilities.

Sometimes you don't have to win the lotto to find Gold – all it really takes is to look within! I'm on my side of the Bridge – I'm Nowhere and I'll leave this one song with you. Band is called Aslan and the song is called "Wish You Were Here."

If you like this book, please leave any comments on Facebook page

The Bridge to Consciousness 'book'.